T0328424

ALSO BY KEVIN RANDLE AND RUSS ESTES:

Faces of the Visitors

OTHER BOOKS BY KEVIN RANDLE:

Conspiracy of Silence

History of UFO Crashes

To Touch the Light

The Truth About the UFO Crash at Roswell

The UFO Casebook

Project Blue Book Exposed

The Randle Report

DOCUMENTARIES BY RUSS ESTES:

The Quality of the Messenger

Roswell Remembered

Magic and Mystery of Giant Rock

The Faces of the Visitors

KEVIN D. RANDLE and RUSS ESTES

A FIRESIDE BOOK / Published by Simon & Schuster

New York   London   Toronto   Sydney   Singapore

SPACESHIPS

OF

THE

VISITORS

An Illustrated Guide
to Alien Spacecraft

FIRESIDE
Rockefeller Center
1230 Avenue of the Americas
New York, NY 10020

Copyright © 2000 by Kevin Randle and Russ Estes
All rights reserved,
including the right of reproduction
in whole or in part in any form.

FIRESIDE and colophon are registered trademarks
of Simon & Schuster, Inc.

Designed by Joseph Rutt

Manufactured in the United States of America

1 2 3 4 5 6 7 8 9 10

Library of Congress Cataloging-in-Publication Data
Randle, Kevin D.
Spaceships of the visitors : an illustrated guide to alien spacecraft /
Kevin D. Randle and Russ Estes
p. cm.
"A Fireside book."
Includes bibliographical references and index.
1. Unidentified flying objects—Sightings and encounters.
I. Estes, Russ.   II. Title.

TL789.R3239   2000
001.942—dc21   00-030863

ISBN 0-684-85739-1

Contents

Introduction

For thousands of years people have been looking into the sky and seeing things they simply did not understand. They labeled them as best they could, calling them flaming swords, fiery chariots, twirling shields, and, eventually, flying saucers. They offered descriptions of these bizarre objects that seem, even today, to defy explanation. These early accounts suggest something that is

beyond the commonplace and the mundane, something that is extraordinary, a phenomenon that began appearing before there was a written language to record observations of it.

In today's world, there have been hundreds of thousands, if not millions, of reports of flying saucers, cigar-shaped craft, glowing spheres, and bright nighttime lights. These reports differ from their older counterparts only in number. We are still confused about what the reports represent and what they may mean. We describe them in the best language available and use the best of the electronic, optical, and unaided observations as we can. And even with all our sophisticated monitoring, sensing, and detection equipment, we are sometimes as confused about what we've seen as our ancestors were.

That is not to say that we haven't been able to learn something about the phenomenon. We have collected hundreds of thousands of reports, as well as thousands of photographs, of those strange things in the sky. We have an advantage because we can record precisely what was seen. Photographs have allowed us to show our fellows what was in the sky that confused, frightened, astounded, and astonished us.

We have been able to use motion pictures to show the movement of these strange things. And when all else failed, we used illustrations and drawings that represented clearly what we saw. These methods have provided us with a record that allows research to be conducted.

To compile this work, we used, literally, hundreds of different sources, ranging from the original reports of the witnesses to the books and magazines that have appeared in the last fifty years. We used the records compiled for Project Blue Book, the official U.S. Air Force investigation. We were also able to use the records originally compiled by the now-defunct civilian inquiries, including those by the National Investigations Committee on Aerial Phenomena (NICAP) and the Aerial Phenomena Research Organization (APRO).

It has been suggested, however, that our view of what the alien spacecraft look like, though once heavily influenced by ancient myths and legends, is now influenced by movies and books. *Close Encounters of the Third Kind,* the spectacular movie that presented an image of alien visitors coming down to Earth, is supposed to have been responsible for our modern view of what alien visitors look like today.

Others have suggested that flying saucers owe their existence to a misuse of the descriptive term by a newspaper reporter. Businessman Kenneth Arnold described the "crescent-shaped" object he had seen as moving with a motion like that of "saucers skipping across the pond." Within days, there were hundreds of sightings of "flying saucers" rather than the crescents seen by Arnold. Without an illustration to accompany the story, people assumed that the flying saucers were disc-shaped.

Of course, as we look at the sighting reports made prior to Arnold's, we note that many did have a disc shape. But we also notice that other shapes seemed to dominate in some eras. The Great Airship stories were not of flying saucers, but of huge, cigar-shaped objects that resembled the airships built on Earth in the years to follow.

The Scandinavian ghosts rockets of 1946 were nearly all shaped like the V-2s that had caused so much fear and destruction at the close of the Second World War. It wasn't until 1947 that these objects began to be reported with any regularity as saucers. However, it must be noted that beginning in the late 1980s, UFOs were being reported as triangular-shaped.

What we see in the survey of the literature, then, is a diversity in the descriptions of alien craft. Although media influence can be seen in some accounts, it also seems that media influences, at least concerning the shape of the craft, have often been ignored. Even when thousands were reporting flying saucers, there were those reporting other, often intriguing shapes. While there is little doubt that the media have influenced the descriptions of alien spaceships, there is also little doubt that many individuals are reporting some craft that were at variance with those media-driven descriptions.

3

We should note here that we are not suggesting that all the stories are science fiction and based on media accounts, or, on the other hand, that all of them are true. These tales are what the witnesses themselves have reported to the air force or the civilian UFO organizations. Those witnesses may have been influenced by what they have seen in the pop culture world around them. They may have been influenced by television and movies and the reports of others. Or, they may not have been influenced by anything other than their own imaginations or by what they saw during their encounters.

We have gathered what we consider to be a representative sample of the visitor-spaceship reports. . . . We have also attempted to tell the story of each encounter in sufficient detail to enable the reader to make an intelligent decision about the validity of the report. And if that is not enough, we have tried to provide a variety of sources where additional information or contrary views can be found.

We have also devised a rating system: We have assigned a "reliability" value to each of the case studies. The rating scale ranges from zero, which means no reliability, to ten, which means that there was actual proof that an alien ship was seen.

It might be useful to note the way that we did assign some of the numbers. There are some truly incredible cases that we rated fairly high. The reason for doing so is the number of witnesses involved in those specific events. A case in which a dozen people report the same thing, and corroborate the tale told by the others, demands that we give it a higher reliability rating because of the large witness count.

On the other hand, there are some cases that are almost universally accepted as reliable that we gave low numbers. Again, it is because of the lack of good corroboration in the form of additional, independent witnesses. The number and the credibility of the witnesses must be taken into account.

But it should be noted that the reliability rating reflects our subjective opinion. We are aware that we tend to believe some of the unbelievable stories over other unbelievable stories. It is a factor that comes from our long work in the field and just how wild some of the tales have become.

What we have here is an encyclopedia of the alien spaceships seen around the world. It is a compilation of the testimonies of hundreds of sources. It is an attempt to inform you about the diversity of the phenomenon and the numbers that have been reported.

So here are the alien craft that have been reported to have visited us in the past. Here are the stories of their visits, based on the testimony of the people who witnessed them. Here is the best analysis we can make of those tales so that you will have a feel for what may just be true and accurate and what is a hoax or a misdirection. And finally, here are the sources of those tales so that you can check them for yourself.

Part I:

ANCIENT HISTORY

From the beginning of time, humankind has kept a record of unusual events. Prehistoric man recorded his history and lore in the form of cave paintings, petroglyphs, and geoglyphs, that is, artificial rock formations. Many of the cave paintings depict creatures and events that could be viewed as sightings of great burning ships in the sky or helmeted alien creatures. Of course they also can be seen as comets, meteors, strange cloud formations, and gods in headdress. When we deal with prehistory, we also deal with prelanguage, and from images without written words to explain them, many conclusions can be drawn.

Some of the best evidence for early alien contact was brought to the public's attention in 1968 with the publication of *Chariots of the Gods?* by Erich Von Daniken. Von Daniken's theory of ancient astronauts is based on early cave drawings and geoglyphs of figures in attire that resembles modern-day space suits and strange helmetlike headgear. Von Daniken's thinking has been influenced by studying ancient roadways and large geoglyphs.

Many large and enigmatic geoglyphs are located on the plain of Nazca in Peru. Some of the glyphs are roadways that look very similar to modern aircraft runways; many are straight-line roads that lead to nowhere. A huge figure more than 820 feet in length is carved into the side of a mountain above the Bay of Pisco, seemingly pointing in the direction of the Plain of Nazca. The archaeologists claim that the giant markings are Inca roads and religious symbols, but Von Daniken sees them as space ports and aerial direction markers. Who is right? We don't know.

There are many unsolved mysteries of the past, and sad to say, most of them will remain mysteries because, once again, we are dealing with prehistory—before discernible record keeping and easy-to-understand language. We must move on to ancient cultures that had language and mathematics that we can understand before we can rate the sightings of unidentified objects.

When we look at early civilizations, we are amazed at their accomplishments: cities that are aligned with the stars, accurate calendars, pyramids, and structural artifacts of gigantic proportion. It is truly a wondrous thing when modern humans look at ancient cultures and marvel.

Or is it a wondrous thing at all? Maybe not!

We tend to underestimate the intelligence and ingenuity of early mankind. We see pyramids that are astronomically aligned, scratch our heads, and wonder, How did they do that? Of course a simple explanation is that if you stack rocks without establishing an internal framework, you get a pyramid. Astronomical alignment has a number of logical explanations. Early humans were hunter-gatherers; they lived by the seasons and documented them well. Logic tells us that if we lived in a time that did not have the diversion of television, radio, or the *Old Farmer's Almanac,* we would spend more time looking to the skies and documenting the movement of celestial bodies. Historians have proven that most early cultures were acutely aware of the movement of the stars and seasonal changes.

In many early civilizations, astronomy was sufficiently advanced to the point that highly reliable calendars had been developed. In China, a calendar had been developed by the twenty-sixth century B.C. Also, a Chinese astronomer, Shih Shen, drew up what might be the earliest star catalog, which listed more than eight hundred stars. Shih Shen's catalog also listed comets, meteors, large sunspots, and supernovas. All of this detailed work in astronomy was done without the benefit of a telescope, which would not be perfected until A.D. 1609.

Humankind has been blessed with brilliant minds since the beginning of time. The first atomic theory was put forward by the Greek philosophers Leucippus and his disciple Democritus, in the fifth century B.C. These men taught that everything is composed of infinitely tiny indivisible particles called atoms.

Dating to 5000 B.C., Mesopotamia is often called "the cradle of

Nearly everything in the night sky fooled ancient people.

civilization" because it is believed to be the oldest of the ancient civilizations. Mesopotamia developed the first known calendars and cities, the earliest known written language, the advancement of commerce and such sciences as astronomy and mathematics. About 2000 B.C., King Hammurabi set down a code of laws that is still the basis for all of the laws in the free world.

In the old testament of the Bible, the prophet Ezekiel relates a sighting of great import: a large structure comprising four sets of sparkling rings, each set a wheel within a wheel. Above the craft, according to some sources Ezekiel saw a burning godhead, "like glowing metal, as if full of fire," surrounded by a brilliant light. He described it as a great cloud with "brightness round about" it and fire flashing forth. From the center, there was gleaming bronze and the "likeness of four living creatures" in the form of man, but

A NASA engineer reproduced Ezekiel's wheel using modern techniques.

each had four faces and four wings. Ezekiel interpreted this sighting as "the likeness of the glory of the Lord."

Ezekiel's wheels could possibly be the first documented sighting of a UFO, or so thought Erich Von Daniken, who stated his belief to that effect in his 1968 book *Chariots of the Gods?*

Upon reading about Ezekiel's biblical vision, Josef F. Blumrich, an engineer with the National Aeronautics and Space Administration (NASA), set out to debunk the Von Daniken theory, using a rocket engineer's rigorous examination. After all, who would know how to investigate the theory better than an engineer who was involved in aircraft design since 1934? Blumrich had played a major role in building NASA's Saturn V rocket, and he was convinced that Ezekiel's wheels would never work. After a great deal of work on the design described in the Bible, Blumrich was surprised and shocked to find that he could adapt it into a practical design for a landing craft. He was so elated with what he found that he worked out the format in detail and published an account of it in a 1973 book titled *The Spaceships of Ezekiel*. A man who set out to disprove Von Daniken's theory did just the opposite. Blumrich

It seems that the object above the city is piloted.

A storm or a manufactured craft?

wrote, "Seldom has a total defeat been so rewarding, so fascinating, and so delightful!"

The date of Ezekiel's sighting was approximately 593 B.C., making this the first UFO sighting that was documented with the written word.

For the next thousand years or so, we slip back into verbal legend or visual images. Many of the biblical scenes painted by fourteenth- and fifteenth-century artists have strange craft in the background. One such piece even has what looks like a rocket piloted by a man.

In the fifteenth century, the invention of the printing press and movable type enabled written and pictorial matter to be reproduced in quantity. From the sixteenth to the eighteenth century, news of all kinds reached the general public by way of illustrated, printed broadsheets. We now had a medium that consisted of written language with pictures, which could document events of the time. It is in just such a broadsheet that we find what could be the first well-documented UFO sighting.

The year was A.D. 1561 in the German city of Nuremberg. A battle in the sky over the city was reported, involving strange craft that were described as red, blue, and black balls; crosses; and tubes. As much as some theorists would like us to think that this was nothing more than a large storm, the fact that many people saw the same thing and all of them described it as colored balls, crosses, and tubes tells us that no storm produced those visions. As history has shown us, it has been thousands of years since humankind thought that a storm was a portent or a vision from God.

In 1566 the citizens of Basel, Switzerland, witnessed a similar display. According to the accounts that were published, the sky was suddenly dotted with large black spheres that were zooming toward the sun and maneuvering around each other. Then as quickly as they arrived, they turned a fiery red and vanished. Basel's story is another well-documented account of strange craft that were witnessed by many and doubted by few.

In the 1680s a French coin was minted, depicting a large saucer-shaped craft over a placid village in broad daylight. Unfortunately there is no printed material that can be found explaining the craft

or the reason that the French government chose to place it on their coin.

The acclaimed British astronomer Edmond Halley not only found and plotted the comet that was named after him, he also had sighted some unexplained aerial objects. It was March of 1716 when Halley spotted a number of objects in the sky. One of them was glowing so brightly that it lit up the sky for more than two hours, and he claimed that he could read written text by its light. He watched as the glow began to wane, and then it suddenly flared up again "as if new fuel had been cast on a fire." If we were to rate this sighting based on the credibility of the witness, we would have to give Edmond Halley a 10.

Were these sightings spaceships of the visitors?

We may never know, but whatever they were, history has shown us that they have been here for a very long time.

Part II:

THE GREAT AIRSHIP

OF 1897

Contrary to popular opinion, the first UFO wave was not in 1947, or even during World War II with its Foo Fighters and glowing balls of gas. It actually began in 1896 and carried over into March and April of 1897. It was a full-blown UFO wave, complete with occupant sightings, abductions, cattle mutilations, landings, and crash retrievals. It foreshadowed all of the modern UFO sightings right up to, and including one or two photographs of, the craft.

It was on the evening of November 17, 1896, that a light like an "electric arc lamp propelled by some mysterious force" passed over Sacramento, California, and touched off the reports of the Great Airship. Once that information had been printed in one of the northern California newspapers, other papers started to make their own announcements and print similar stories.

Almost from the beginning, the *San Francisco Chronicle* claimed that the Sacramento airship was a hoax but also noted that those who had made reports were sincere. On November 20, just three days after the sighting, the San Francisco newspaper quoted Professor George Davidson, who believed the whole episode was the result of "a sort of free masonry of liars. Half a dozen fellows got together, sent up a balloon with some sort of an electric light attached, and imagination has done the rest. It is a pure fake."

Airship described by eyewitnesses as it passed over Sacramento, California, in November 1896.

Even so, the airship returned to Sacramento on November 20 and then appeared over the San Francisco area on November 21. Passengers on a streetcar in Oakland saw something that was described as a "peculiar-looking contrivance."

As one San Francisco newspaper, the *Examiner*, was reporting these events, the *Chronicle* was suggesting that a lawyer, George D. Collins, had met with the inventor of the airship. Collins, according to the *Chronicle*, had seen the machine which he described as being 150 feet long with two canvas wings and a rudder like the tail of a bird. The airship had been built in Oroville, California, by a wealthy man who had moved there from Maine. As soon as the inventor resolved some minor technical problems, he would fly the ship over the city so that everyone could get a good look at it.

The *Examiner* sent a reporter from San Francisco to Oroville, but the man could find no one in the town who knew of the inventor or of a wealthy man who had moved there from Maine or of the airship. Without any sort of corroboration for the story, he concluded that it was nothing more than a hoax. That was supposed to signal the end of the airship.

To make matters worse, Collins approached still another California newspaper to tell them that the *Examiner* had misquoted him. The quotes about the airship had come from a friend of Collins's whom the *Examiner* had interviewed. Collins, who had not spoken to anyone at the *Examiner*, said that everything his friend had told the reporter was a lie.

Confusion reigned as suspicion fell upon a dentist as the suspected airship inventor. He had come from Carmel, Maine, and spent time tinkering with inventions. He also seemed to fit the profile of the mysterious airship inventor. It was confirmed that his attorney was Collins, but he claimed that all his inventions had to do with dentistry. The dentist, sick of all the attention, hid out somewhere, and reporters, searching for a story, illegally rifled through his possessions, finding nothing other than copper dental fillings.

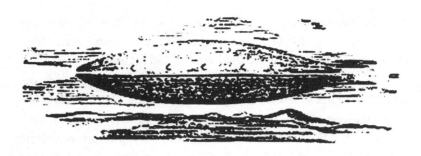

Airship as described in California in November 1896

Once it was believed that the identity of the inventor and the location of his invention had been discovered, the story began to fade, only to be revitalized almost immediately. On November 24, 1896, the *Oakland Tribune* reported that George Carleton, a local resident, knew the inventor, or at the least, knew his name. Of course, Carleton was sworn to secrecy and wouldn't expose the inventor. All he would do was confirm that the craft was being tested in the Oroville area, which fit nicely with all the other stories that had been reported previously.

W.H.H. Hart, onetime attorney general of California, announced that the inventor, irritated with Collins for talking too much, had fired Collins and hired Hart. According to Hart, there were *two* airships, and his job was to consolidate the interests. Hart seemed to believe that one of the airships would be used for war, and there was talk of dropping dynamite on Havana, Cuba, where the Spanish were causing a great deal of trouble.

Hart's confirmation of the airship, and a "secret informer's" new information that the ship had been flown from Oroville to Sacramento to a barn, seemed to prove the existence of the airship to some. It had been flown every night, according to various sources, and that was exactly what people had been reporting. The "voices" heard by some on the ground were identified as engine noise rather than human speech.

Hart, however, began to retract a number of his earlier statements, suggesting that he had never actually seen the airship himself. He had talked to a man who had claimed to be the inventor but had no proof that such was the case. He had been convinced of the reality of the airship by the unidentified inventor.

Hart's retractions didn't end the sightings, however. They continued to be made all along the California coast. While some of the sightings were little more than nocturnal lights, that is, strange lights seen in the distance late at night, other sightings were more detailed. On November 26, 1896, for example, Case Gilson claimed that he had seen an unlighted airship in the sky above Oakland. Gilson said that it "looked like a great black cigar with a fishlike tail. . . . The body was at least 100 feet long and attached to it was a triangular tail. . . . The surface of the airship looked as if it were made of aluminum. . . ."

Others reported they had seen the airship on the ground. John A. Horen told San Francisco's *Examiner* that he had met a stranger who took him to a remote location where the two of them boarded the airship. Horen was then treated to a trip to Hawaii and back, made in a single night. Horen's wife, however, found the tale ridiculous, saying that he had been at home sound asleep, next to her, when the flight had allegedly taken place.

On December 2, fishermen near Pacific Grove, California, supposedly saw an airship land on the water and then float slowly to shore. It was occupied by three men, one of whom was addressed as "Captain" by the others. The fishermen were told that the ship required some repairs and that the captain was not yet ready to announce his invention to the rest of the world.

After that, still more people came forward with tales of meeting with the crews of those piloting the airship. In most cases, the airship had some kind of mechanical problems requiring it to land for a short period for repairs. George Jennings told reporters that a traveler had entered his business—a man Jennings claimed to have recognized but whom he would not identify—who was one of the airship inventors. Jennings said he knew the man well and there had been no reason for him to lie about his invention.

What might be the first recorded claim of an attempted alien abduction associated with the airship stories came during the early evening of November 25, 1896. Colonel H. G. Shaw said he and a companion, Camille Spooner, left Lodi, California, "when the horse stopped suddenly and gave a snort of terror."

Shaw claimed that he saw three figures who stood nearly seven feet tall and were very thin. They looked human and didn't seem to be hostile, so Shaw tried to communicate with them. According to Shaw, they didn't seem to understand him and responded with a "warbling" type of monotone chant.

Shaw continued his description in the newspapers, saying, "They were without any sort of clothing, but were covered with a natural growth as soft as silk to the touch and their skin was like velvet. Their faces and heads were without hair, the ears were very small, and the nose had the appearance of polished ivory, while the eyes were large and lustrous. The mouth, however, was small and it seemed . . . they were without teeth."

Shaw also said they had small nailless hands and long narrow feet. By touching one of the creatures, he discovered that they were nearly weightless. Shaw said he believed they weighed less than an ounce.

With Shaw so close to the beings, they tried to lift him with the intention of carrying him away. When they couldn't budge either Shaw or his companion, they gave up and flashed lights at a nearby bridge where a large "airship" was hovering. They walked with a swaying motion toward the craft, only touching the ground every fifteen feet or so. Then, according to Shaw, "with a little spring they rose to the machine, opened a door in the side and disappeared. . . ."

A week or so later, in early December 1896, two fishermen, Giuseppe Valinziano and Luigi Valdivia, said they had been held captive for a number of hours while the airship crew made repairs. The "captain" of the craft would only provide vague clues about the origin of the ship but did say the invention would be announced to the world within weeks. When the repairs were completed, the men were allowed to leave unharmed.

The California airship arrived in Nebraska, February 1897.

By the middle of December 1896, the airship stories began to fade once again from the California newspapers. On January 16, 1897, however, former attorney general Hart reappeared to say that the airship inventor had left California and was on his way to Cuba where war between the United States and Spain was brewing. Apparently the airship inventor, a patriotic American, thought his invention could be of some use against the Spanish.

In February the airship craze began to spread from California. The *Omaha Daily Bee* reported on airship sightings in the south-central part of Nebraska, which had begun to occur in the fall of 1896. The airship had been seen about five hundred feet above the ground as it hovered for about thirty minutes, according to those early reports.

Near Big Springs, Nebraska, on February 17, three men reported a "barrel-sized light" rising into the air and saw it descend rapidly as it shot out sparks. On February 18, the *Kearney Hub* reported that the "now famous California airship inventor is now in our vicinity."

Late in the month, February 26, a group of people at a railroad depot in Falls City, Nebraska, saw a "big searchlight, moving in a westerly direction, apparently at a speed of about sixty miles an hour, and in the same portion of the sky a red light, much like a [railroad] switch lamp, was plainly seen."

The light seemed to be moving toward Stella, Nebraska, and railroad dispatcher Ike Chidsey, wired the agent there. Within minutes the light was seen over Stella. Other reports were made from Beatrice, Wymore, Hastings, Kenesaw, and Hartwell, all in Nebraska. Some of the witnesses said that they had seen the light for several nights but had been reluctant to report anything for fear of ridicule. Those reports, coming from railroad dispatchers and telegraphers, would foreshadow a much larger and more complex wave of sightings that would begin in a few weeks and that would be reported throughout the Midwest.

With its searchlight beaming, the airship hovered over a number of cities in March and April 1897.

The wave seemed to explode on March 29, 1897, when an airship was again reported over Omaha, Nebraska. At the same time, another airship was reported chasing a farmer near Sioux City, Iowa. Robert Hubbert said that he was riding his bicycle and hoping to see the airship "that the whole country [was] talking about." An anchor was being dragged along the ground by the airship, and it grabbed Hubbert, hauling him from his bike. Suddenly, and "none too soon" his pants ripped and Hubbert fell back to the ground. Although he was physically unhurt, he was angry. He told reporters that it was criminal "for the skipper of the ship to let a grapnel drag on the ground."

The very next night, March 30, the people of Denver, Colorado, reported an airship overhead, and on April 1, it was seen above Kansas City, Kansas. Hundreds reported it as it paused from time to time to play with its searchlight among the clouds. About a half an hour after it disappeared from Kansas City, it was reported over Everest, Kansas. Witnesses there said that it resembled a cigar with wings and that it glowed brightly while hovering.

On April 2, it was seen near Decatur, Michigan. According to the story, the first evidence of the airship was a bright light and then behind it was a dark shape. The witnesses said that they could hear a sharp crackling sound and voices.

On April 10, an airship was over Illinois. From Chicago, thousands of people watched an airship displaying its lights. Later the same evening the airship made at least one landing. As it descended near Calinville, Illinois, a crowd began to gather. It settled into a pasture, and the curious started forward. Apparently the crew thought the townspeople were too close, and the ship took off abruptly.

Also on April 10, an airship was seen over several Iowa towns. Clinton, along the Mississippi River reported it first, then Ottumwa and Albia. That was the third time that the residents of Albia had seen the ship. They were also credited with the first report from Iowa.

On April 12, the airship landed once again. According to the

The railroad employees join the search for the Great Airship.

witnesses, the object was large, cigar-shaped, with wings and a canopy over the top. A man climbed out and walked around as if looking for damage. After fifteen minutes, the airship "rose to great height" and disappeared to the north.

On that same evening, an engineer on a train near Chicago said he watched an airship for several minutes but "was forced to turn my attention back to my duties." When he looked up again, the airship was far ahead of the train, and near Lisle, Illinois, he lost sight of it.

The middle of April brought the reports of a series of landings in Iowa. The *Cedar Rapids Evening Gazette* reported that an airship had landed on the Union Station in the "wee morning" hours and that several local citizens were taken on board. Charley Jordan quickly made his story known and even signed an affidavit attesting to the truthfulness of his tale of his flight. He was described as "never telling but a few lies and then only about things of impor-

27

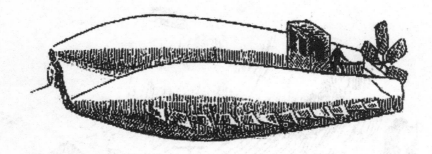

The Waterloo Airship was an admitted hoax.

tance." Also taken for a flight, W. R. Boyd said his whole purpose in going was to "get as high as possible so that he could learn about the condition of the post office." The members of the strange crew were reported to be tired from their journey but promised to lecture about their trip quite soon. The topics to be discussed included the unlikely subject of hell.

The night after the Cedar Rapids report, the airship was captured in Waterloo, Iowa. The *Waterloo Courier* reported that the unusual craft "came to rest on the fair ground" and one of the pilots went to the police station to ask that they guard his ship. Arriving at the fairgrounds, the police found a large, twin-cigar-shaped object. All during the day people came to see the ship, the first tangible object to be found that didn't disappear into the night sky. That made the airship stories a little more plausible.

A "professor," who spoke with a heavy accent and claimed to come from San Francisco, told of the dangerous flight across the country that ended in tragedy when the leader of the expedition fell into the Cedar River. Attempts to rescue the man failed.

By late afternoon, interest in his story was waning, and then ended abruptly. The professor was recognized as a local man, E. A. Feather. He dropped his accent and the ship was finally removed from the fairgrounds, but not before hundreds had seen it and more than one newspaper article had been written about it.

An airship was also making the rounds in Texas during April.

A man from Denton, north of Dallas, said that he had seen the object and it was definitely some "kind of manufactured craft." From Hillsboro, Texas, came the report of a "brilliant light, as if coming from an arc light . . ." and then it was seen gliding over a field near by.

Early on the morning of April 17, two men from Rhome, Texas, said that they saw the airship heading west at 150 miles an hour. The same day, the *Fort Worth Register,* which "hardly cares to repeat it," reported that a man traveling near Cisco, Texas, saw the airship landed in a field. Several men were standing around the craft, and Patrick Barnes went over to talk to them. At the ship he was told by the crew that they had some kind of engine trouble but would be leaving soon to go to Cuba to "bomb the Spanish." By one o'clock they had repaired the craft, and they took off for the Ozarks to train for their mission.

In Paris, Texas, a night watchman said that he saw a cigar-shaped craft, two hundred feet long, with large wings. Later, in Farmersville, several people said they heard the crew of an airship singing.

April 17, 1897, might be said to be the high point of the airship stories. The most spectacular, best-known, and most widely reported of the tales came from Aurora, Texas, on that day. According to the newspaper reports, at about dawn, an airship appeared on the horizon south of the town. Dozens in Aurora watched as it came in low, buzzed the town square, and then continued on to the north. On Judge Proctor's farm, it struck a windmill and exploded. Proctor's house and flower beds were damaged as the airship disintegrated. Dozens of people ran to the scene. In the mangled wreckage they found the body of the pilot badly disfigured. According to the early reports, T. J. Weems, a U. S. Signal Corps (intelligence) officer, said that the poor dead creature probably came from Mars.

Searchers also found several documents covered in a strange writing. They managed to decipher it, learning that the airship weighed several tons and was made of silver and aluminum. By noon, however, all the debris had been cleaned up, and late in the day, the "Martian" was given a Christian burial in the local cemetery.

It must be noted here that there is no good evidence that the story is true. Those mentioned in the original stories published in 1897 are not who they were described to be. T. J. Weems, for example, was not, as printed, a member of the army's Signal Corps but was, in fact, the local blacksmith. Claiming that he was a Signal Corps officer was an attempt to lend additional credibility to the story.

Longtime residents of the area, interviewed by Kevin Randle in 1972, including members of the Wise County Historical Society (in Aurora, Texas), said that the explosion and crash hadn't happened. Records suggested that there was a Judge Proctor and he did own land in the area in 1897. However, according to those who lived there, Proctor did not have a windmill. (At least the landowner in 1969 said that, but later with all the publicity about the crash, he claimed that Proctor had two wells, including one with a windmill.)

The Aurora story originated with H. E. Hayden, a stringer for the *Dallas Morning News* in 1897. According to some sources, Hayden admitted to inventing the tale to put his hometown back on the map. In 1897, after a disastrous outbreak of disease, and a decision by the railroad to by-pass Aurora, the town was dying. Hayden wanted some way of promoting its existence (not a bad idea when the fame of Roswell, New Mexico, is considered today.)

The Aurora crash wasn't the only destruction of an airship reported. In late April 1897 a longtime resident of San Angelo, Texas, claimed he saw an airship fly into a flock of birds and explode. He was unable to find the wreckage and had nothing to prove his story.

On the other hand, other airships were still intact. On April 22, in Josserand, Texas, Frank Nichols was awakened by a whirring

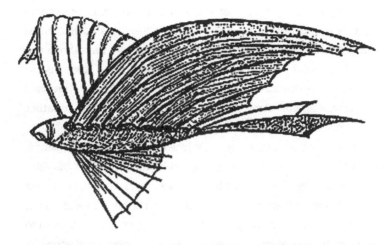

One of the strangest of the airships, reported near Dallas, Texas, in April 1897.

noise like that of machinery. On the ground, in his cornfield, was an airship, and near it were two men with buckets. They asked Nichols for permission to use his well. Before they took off, they told Nichols they were from Iowa, and in a few weeks, they would reveal their invention.

The day before, according to an affidavit signed by a number of men in and around LeRoy, Kansas, the airship had been hovering over the ranch owned by Alexander Hamilton. Hamilton, in April 1897, reported that one of his cows had been stolen from a closed corral late at night. The entire account of the aerial cattle rustling was published in the *Yates Center Farmer's Advocate* on April 23. Hamilton and a number of his friends signed an affidavit attesting to the truthfulness of the story.

> Last Monday night [April 19] about half past ten we were awakened by a noise among the cattle. I arose thinking perhaps my bulldog was performing some pranks, but upon going to the door, saw to my utter amazement, an airship slowly descending over my cow lot about 40 rods from the house.

Calling Gid Heslip, my tenant, and my son, Wall, we seized some axes and ran to the corral. Meanwhile the ship had been gently descending until it was not more than 30 feet above the ground and we came up to within 50 yards of it. . . . It was occupied by six of the strangest beings I ever saw. There were two men, a woman and three children. They were jabbering together but we could not understand. . . .

When about 30 feet above us, it [the airship] seemed to pause, and hover directly over a three-year-old heifer which was bawling and jumping, apparently fast in the fence. Going to her, we found a cable about half an inch in thickness . . . fastened in a slip knot around her neck, one end passing up to the vessel. . . . We tried to get it off but could not, so we cut the wire loose, and stood in amazement to see the ship, cow and all rise slowly and sail off. . . .

Link Thomas, who lives in Coffee County about three or four miles west of LeRoy, had found the hide, legs, and head in his field that day. He, thinking someone had butchered a stolen beast and thrown the hide away, had brought it to town for identification but was greatly mystified in not being able to find a track of any kind on the soft ground.

The affidavit was signed by a number of men who claimed they had known Hamilton for years, stating, "that for truth and veracity we have never heard his word questioned and that we do verily believe his statement to be true and correct."

During the mid-1960s, the Hamilton tale surfaced again and was repeated in a number of magazine articles and UFO books. Each time, the statement of Hamilton's friends, attesting to his veracity, was mentioned without question. Here was a tale that deviated from the airship stories in a number of ways, suggested an extraterrestrial explanation for the event, and involved the rustling of a cow. It proved the strangeness of some of the airship tales and suggested to those who wanted to believe in them that something otherworldly was happening.

UFO Encyclopedia author Jerome "Jerry" Clark, who had reported the Hamilton case seriously a number of times, did some additional investigation on his own in the 1970s. He learned that one of Hamilton's daughters still lived in Kansas and interviewed her about the story. Although she hadn't been born until after the event, she had heard her father talk about it on many occasions. She said that Hamilton belonged to a local liars club, as did all the men who signed the affidavit attesting to Hamilton's veracity. It suggested a motive for the tale that had nothing to do with extraterrestrial visitation.

Even worse was a letter found by folklorist Dr. Thomas "Eddie" Bullard. In the May 7, 1897, edition of the *Atchison County Mail*, Hamilton admitted that he had fabricated the tale. There simply is no reason for it to continue to circulate as true. The evidence that it was a lie has been uncovered and reported.

Hamilton's tale, one of the most widely reported and accepted of the airship stories, was also one of the few that suggested an outer-space connection rather than an earthly tradition. Most of those who reported seeing or communicating with the pilots and crews suggested they were fellow Americans, some just trying out their new invention and others with a mission to Cuba to attack the Spanish. The airship was an earthly invention with no outer-space connection.

One tale of extraterrestrial visitation, however, came from Reynolds, Michigan, on April 14. According to the witnesses, a flying machine, as opposed to an airship, came down half a mile from town. A dozen farmers who had watched it maneuvering overhead rushed to the landing site. Inside was a giant, manlike creature, whose speech was described as musical. Although there were what looked to be polar-bear hides inside the cabin, the alien creature didn't use them and seemed to be suffering from the heat.

One of the farmers tried to approach the ship but was kicked by the giant, resulting in a broken hip for the farmer. Although the pilot was unfriendly, the ship apparently didn't take off, and people from surrounding towns hurried to the scene. The creature tried to talk to the people but eventually gave up.

On April 15, near Linn Grove, Iowa, an airship appeared overhead. Five local men followed it for four miles where it landed in the country. When the men got close to the ship, it spread four massive wings and lifted off. The occupants threw out two boulders of an unknown material. The occupants were reported as "queer-looking beings with extraordinarily long beards."

In another report that would foreshadow much of the UFO sightings in the modern era, W. H. Hopkins claimed that he had seen one of the airships sitting on the ground in a clearing. Next to it, Hopkins said, was the "most beautiful being I ever beheld." She was naked, with hair that hung to her waist. She was picking flowers and had a lovely, musical voice. Lying in the shadow of the ship was a naked man, with a long beard and long hair, who was fanning himself as if warm.

As Hopkins approached, the woman screamed and ran to the man. Hopkins attempted to tell them that he meant them no harm, and eventually communication was established. He asked where they came from, and they pointed at the sky, "pronouncing a word which, to [Hopkin's] imagination, sounded like Mars."

Hopkins was given a tour of the ship. The two beings seemed interested in his clothes and his gray hair. They also examined his watch "with the greatest wonder." Finally he exited the ship and they flew away, "laughing and waving."

In Merkel, Texas, not all that far from Aurora, dozens of people returning from church saw a ship, its anchor snagged in a barbed-wire fence. Climbing down a rope in a blue sailor suit was a small man. When he spotted the people, he scurried back up into the airship. The people cut the anchor free and kept it on display for several weeks. It has, quite naturally, disappeared.

There were additional sightings in May. As a single example, there is the report from near Cassville, Indiana, on May 3. Edwin Shaffer, while passing a gravel pit, noticed a cigar-shaped, forty-foot-long airship sitting in it. He claimed that it was "handsomely furnished on the inside and the aerial craft was inhabited by a crew of midgets who spoke no English."

But the intense interest in the airship was dying out and by the

34

summer of 1897, there were almost no new tales. By the time the modern-era flying saucers began to be reported fifty years later, almost no one remembered the airships. UFO researchers did not begin to dig for the stories until the 1960s, though there had been a few discussions about them earlier. Captain Edward Ruppelt, at one time the chief of Project Blue Book, the official U. S. Air Force investigation, mentioned the wave briefly in his book *The Report on Unidentified Flying Objects.* Others, such as Charles Fort, chronicler of the bizarre, devoted a little space to the reports in books about the unusual, commenting on their strange nature but not examining the stories with anything close to a critical eye.

It wasn't until the 1960s that the airship wave was rediscovered and the reports began to enter the UFO literature. Jacques Vallee discovered the Alexander Hamilton "calfnapping" and reported it in his *Anatomy of a Phenomenon.* It was then quoted in a large number of books, arguing for the reality of UFOs in general and an extraterrestrial hypothesis in particular.

Vallee's discovery and the notations made by Fort and others inspired many to search old newspaper records. Hundreds of accounts were discovered. These early airship sighting were sometimes reported as straight news, while other sighting articles had a tongue-in-cheek quality that is often missing in modern newspapers. The latter also suggested that while the stories were being told, the reporters didn't accept them at face value.

The problem was that those doing the research in the 1960s made an invalid assumption. They believed that what had been printed in the newspapers in 1897 was the truth, the whole truth, and nothing but the truth. The reality was that many of the airship accounts were fabricated, just as some of the "straight" news accounts of wars, disease, and other disasters were invented by reporters and editors to sell the newspapers. The public was interested in the airship, so reporters and editors gave them what they wanted. They published tales of the airship overhead and speculations about its origins and its mission.

The media-hype factor is borne out in a tale from Burlington,

Iowa, that was described, at the time as one of the "meanest and most discouraging stories of the entire lot." Members of the Burlington newspaper staff sent up a common tissue-paper hot-air balloon so that it would fly over the city. The staff began to get reports of the airship, which they then dutifully reported in the newspaper. One of the most distinguished men in the town came forward to say that he had not only seen the airship but had heard voices from it, and that he would even sign an affidavit about it. That convinced the Burlington reporters that the entire airship episode was a fake.

Some newspapers from 1897 published letters from outraged citizens who claimed they had been identified as witnesses to seeing the airship, and they wrote that they had never seen it. Jerry Clark reported that J. H. Tibbles of Rochelle, Illinois, "wrote the *Chicago Record* that a report in a Chicago paper notwithstanding, no one in Rochelle had seen an airship on April 3; 'I took it upon myself to hunt down the report, and for several days I have been busy doing so. . . . I have not found a person who had seen another person who claimed to have seen it.'"

Eddie Bullard, the folklorist, noted "how seldom airship reports turned up in the columns devoted to news from outlying communities. The content of this correspondence includes sicknesses, births, deaths, marriages, crop news, and mention of anything new or changing in these areas where newness or change was rare. A crime was a major event, and an airship sighting would surely rate a mention. After reading about a lot of Sunday picnics, weekend visits and fine hunting dogs, I can say with safety that those mentions are lacking."

In other cases it's clear that the reporters and editors didn't believe the stories, though they were willing to report them. The airship story that originated in Cedar Rapids, Iowa, was reported as straight news, but it is apparent that the reporters and the editors didn't take it seriously. They did, however, report the story right up to the point when the Cedar Rapids tale was trumped by the Waterloo story. Remember, in Waterloo, they had captured the ship. And that story was reported as straight news during the first days, until the "professor" was recognized, and the joke was revealed.

Bullard's research corroborates his theory. He was suggesting that the reporters and stringers in the outlying areas were more reliable than those in the city. He was suggesting that the newspapers of the day were making it up as they went along. The story from Burlington, Iowa, certainly suggests this is true.

Jerry Clark, who has studied the airship-sighting wave in great detail, noted that by the mid-to-late 1970s "it was becoming increasingly evident that hoaxes played a far more prominent role than anyone had imagined." He was suggesting, that unlike the modern era in which the hoaxes are believed to be a relatively minor fraction of the whole, in 1897, a great number of the reports, including those that had seemed to be the most reliable, were found to be hoaxes.

As mentioned, Alexander Hamilton had invented the tale of the calfnapping, and his friends who signed the affidavit were fellow members of a liars club. The Linn Grove landing had never happened according to a man who lived there in 1897 and remembered nothing about it. The Aurora, Texas, crash was more fantasy, written by a man who wanted his town back on the map. And the Merkel, Texas, anchor-dragging airship was just another of the hoaxes that filled the newspapers.

But there were others who jumped on the bandwagon in 1897. The *Des Moines Register* in 1897 put forth another theory about the reliability and the genesis of the airship, at least in Iowa. The reporters noted that an airship was mentioned in Cedar Rapids on April 14, and on the next night it was seen near Fairfield. It was also seen near Evanston, Illinois, "worrying the Chicago papers greatly." The most remarkable account of the airship came on April 15 from an area near Pella, Iowa. According to the *Register*, "many people, among them the Western Union operator, had seen the machine . . . if it was true, the Pella airship looked like a sea serpent, a balloon, a winged cigar, a pair of balloons hitched together with a car swung between them, a car with an aeroplane and three sails, and 19 or 20 other things."

The *Register* article continued by reporting that the telephone at the *Leader* (another Iowa newspaper) rang and the town of Stuart was "found to be clamoring for fame." They had seen the airship, too. The story went out over the wire and the Pella Western Union operator said that he could produce dozens of witnesses if anyone cared. He said that the airship had come from the southeast, was traveling about fifteen miles an hour, and had a red light in front and a green one in the rear. The operator's feelings were hurt when he was asked if it was an April Fools' joke.

While the conversation between the *Register's* reporters and the Western Union telegrapher was evolving into a heated argument, a report came in that the airship was now over Panora, Iowa. The Western Union operator there said that they had seen the airship over their own town, coming from the direction of Stuart. It was now moving faster but had the same appearance as it did in Stuart, which the *Register* labeled as a "neat attempt at getting around the description."

As the argument increased in intensity, the number of telegrams about the airship also increased. From Clinton, Iowa, came a telegram saying an airship had flown over the town on April 10. Although the airship was reported to have been seen by several reputable citizens, the telegram was almost apologetic.

Immediately came a telegram from Ottumwa, reporting the residents there had seen an airship more than once. "An Eldon [Iowa] operator discovered the airship at 7:25 p.m. Ottumwa was prepared for its appearance. It was seen here by half the population. All agreed that it appeared as a red light moving up and down and traveling northwest. Albia caught sight of it at 8:10 and at 9 o'clock it was still visible. . . . This was the third time that it has been seen in Albia."

The *Register* reported, "The fact seems to be that the airship has been exploited beautifully by telegraphers along certain lines of the railroad. They managed it beautifully for awhile and never allowed it to travel too far too fast." The reports were always well done showing a certain amount of genius. But the rest of the public began to get involved with the sightings and the airship reports got too numerous. Some of these would conflict, and it became evident that someone would have to have a whole family of airships for all the sightings to be true.

What all these accounts suggest is that the vast majority of the airship stories were hoaxes. Some were originated by individuals such as Alexander Hamilton or the people in Cedar Rapids, others were initiated by the newspapers looking for something spectacular to report, and the last bunch were created by the telegraphers along the railroads who were bored late in the evening.

It is now clear that there was no great airship invention just before the turn of the last century. Heavier-than-air flight would become a reality in six years. Airplanes would soon begin flying across the country. And there would be airships built by the military to search for enemy subs or to hover above battlefields so that generals could gather intelligence about enemy movements.

An airship built near Oakland, California, in 1912.

Eventually there would even be airship flights across the Atlantic. These would end when the *Hindenburg* exploded in Lakehurst, New Jersey, in 1937.

But there is no evidence that a human inventor had flown a Great Airship in 1896 or 1897 anywhere in the country. Although some stories suggested the announcement about the airships was about to be made to the world, it never was. And, those alleged airship crewman on their way to Cuba to bomb the Spanish never made it to drop any of those bombs.

A few modern investigators have suggested that there was a solid core of airship sightings, so something had to trigger the tales in 1896. Those researchers have recommended that we examine more closely the stories told in the Sacramento area in November 1896. Such research might provide a clue as to where and why these stories began to circulate.

But even if there was some event that triggered the Sacramento reports, and even if we could identify the core of solid sightings, there isn't much to learn from the data today. There are no witnesses left to tell us what they really saw in 1896 and 1897. Few have interviewed anyone who was around in 1897 and who claimed to have seen the Great Airship. Ed Ruppelt, while chief of the air force's Project Blue Book, wrote that he had had a long conversation with a man who had been a copy boy at the *San Francisco Chronicle* in the time of the airship. He remembered almost nothing about those long-ago events except to tell Ruppelt that the editors and a few others at the newspaper had seen the airship.

But even if Ruppelt's interviewee's statements were true, there were so many tales invented by newspaper editors and reporters that a single, fading memory of a second-hand report means very little today. Maybe something unusual was seen near San Francisco or Sacramento. Maybe there had been some kind of cigar-shaped object over California long ago. Those who saw it did the best they could to describe it, using the terminology available at the time. Maybe there was a sighting or two of something that was not invention, imagination, delusion, misidentification, or outright fabrication in the fall of 1896.

What we know today is that the vast majority of the airship cases can be explained as hoaxes, but they shouldn't be completely ignored. They provide us with an insight that will help us better understand the UFO situation as it stands today.

Part III:

THE PROJECT BLUE BOOK YEARS

About fifty years after the Great Airship had disappeared, strange things were again seen in the sky above the North American continent. There had been tales of mysterious lights and weird objects during those intervening fifty years, but they gained little in the way of national or international interest. There might be a mention in a local newspaper, and if Charles Fort had been alive, he might have recorded them, but for the most part, the sightings had been ignored. Then came the Foo Fighters that harassed Allied and Axis combatants during the Second World War.

Although these accounts were rare in the first years of the war, there were a few. Typical of those early sightings was the report from a marine with the First Marine Division who had invaded the island of Tulagi, just west of Guadalcanal in August 1942. According to the story, reported most recently by Jerry Clark in his massive *UFO Encyclopedia*, Sergeant Stephen J. Brickner had been cleaning his rifle when the air-raid siren sounded. Brickner took cover and heard the sound of a formation of aircraft of some kind overhead. It didn't sound like the engines of Japanese bombers, and when the objects appeared, they were silvery, flying higher, and in far greater numbers than expected of the enemy. No bombs were dropped, and there was no attack. The objects eventually disappeared in the distance. Brinkner had no expla-

Typical of the Foo Fighters are these lights separating two fighters.

nation for the sighting, and it was listed as an example of the Foo Fighters.

Later in the war, more such stories were told by aircrews, especially those in the European theater of operations. In 1943, for example, Sergeant Louis Kiss, a B-17 tail gunner, reported that a basketball-shaped object, in a shimmering gold color, approached his aircraft from behind and below. It climbed and then seemed to hover over one wing and then crossed over the top of the aircraft to hover over the other, before it disappeared in the rear.

A Royal Air Force bomber on a mission to Germany was paced by a luminous orange globe or disc for several minutes. The tail gunner, who first sighted the object, reported it to the pilot, who also saw it. As it closed to within one or two hundred yards, the gunner fired on it. He believed that he had hit it, but there was no apparent effect. The object finally disappeared in a burst of speed estimated at more than a thousand miles an hour.

Allied intelligence was afraid that the Foo Fighters were a German secret weapon of some sort. They were concerned by the reported performance, maneuverability, and the speed of the Foo Fighters. If it had been an enemy weapon that was soon to be deployed, it could have changed the course of the war, given the superiority of the Foo Fighters over Allied fighters and bombers.

When the war ended, Allied intelligence learned that enemy pilots had also reported the Foo Fighters through their command structure to Axis intelligence, who believed the weapons to be of Allied design. In other words, Axis intelligence was as confused by the sightings as were the Allies.

Questions about the nature of the Foo Fighters were then raised. In 1952 Albert Rosenthal, a soldier during the war, noted that the Foo Fighters were sometimes associated with high concentrations of enemy antiaircraft fire and that they sometimes exploded when chased. He wrote that the Allies never solved the problem of what they were but suggested they might have been German barrage balloons, a secret weapon, or a form of Saint Elmo's fire.

The suggestion of Saint Elmo's fire (a type of static electricity) took on added weight when some of the stories of Foo Fighters were examined carefully. The Foo Fighters, in some cases, seemed to be attached to aircraft, causing ground-based witnesses to believe the plane was on fire. In other cases, aircraft flew through the Foo Fighters, the crews watching as the wings or engines cut through the lights with no effect on either the aircraft or the light.

During the war, it appears that some sort of official investigation was conducted, though little has been reported about that work. In 1953, the CIA sponsored a panel to review the best of the UFO material. Headed by Dr. H. P. Robertson, it made a brief reference to the Foo Fighters, stating that they were "believed to be electrostatic (similar to Saint Elmo's fire) or electromagnetic phenomena or possibly light reflections from ice crystals in the air, but their exact cause or nature was never defined."

Less than a year after the war ended in Europe, there was a series of sightings of what would become known as the ghost rock-

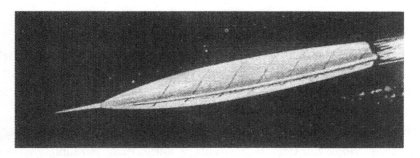

The typical ghost rocket resembled the weapons used by the Germans at the close of the Second World War.

ets. The first of these were reported in the skies of Helsinki on February 26, 1946. A radio station noted that "numerous meteors have recently fallen" in northern Finland.

On June 9, the question about the identity of the objects surfaced again in Helsinki, where a glowing body, trailing smoke behind it, was seen heading to the southwest at a low altitude.

On June 10, witnesses reported something resembling the German V-2 rockets of the Second World War overhead. It was in sight for about ten minutes, and about two minutes after it disappeared, an explosion was heard.

And then, over the next several weeks, there would be other reports of ghost rockets. In many of the accounts, it was reported that the object, or rocket, had crashed. On July 9, 1946, for example, near Lake Barken, Sweden, a witness watched an object with alternating blue and green lights come from the northeast and plunge into the lake about 110 yards away.

On July 10, 1946, at Bjorkon, Sweden, a number of people watched as a "projectile trailing luminous smoke" slammed into a beach, leaving a yard-wide shallow crater containing a slaglike material, some of it reduced to powder. A newspaper reporter found a cylinder about twenty or thirty meters in diameter. Military authorities investigated, produced ambiguous results, and finally accused the witnesses of imagining things.

Just over a week later, on July 18, 1946, at Lake Mjosa, Sweden, two eight-foot-long missiles with wings set about three feet from their front, plunged into the lake, creating "notable turbulence." While in flight, the wings seemed to flap, as if made of cloth, and the objects whistled as they flew.

The next day, at Lake Kolmjarv, Sweden, witnesses watched a gray rocket-shaped object with wings crash into the lake, sparking a three-week hunt for it by military authorities. Nothing was found. Nearly forty years later, a Swedish UFO researcher, Clas Svahn, interviewed some of the civilian witnesses and military investigators. An air force officer speculated that the object might have been made of a lightweight material that could disintegrate easily. A civilian witness claimed she heard a "thunderclap" that might have been the object exploding. There were several additional witnesses to this, and a governmental investigation team was dispatched. All the information about this case would not be available until the Swedish government released the details of their investigation into the ghost rockets much later.

The *Chicago Tribune,* on July 27, reported that there had been more than five hundred sightings of the ghost rockets in just twenty-six days. It was about that time that the Swedish government began to censor the reports. The press could continue to report specific incidents as long as the names of the locations were deleted. It is also important to note that the names of the witnesses were now being left out of the accounts. That, of course, prevented anyone from doing any follow-up investigation and learning things that the military and the Swedish government might not want them to learn.

About the beginning of August, the Swedish Defense Staff announced that "radio-controlled rockets" were entering Sweden from the south and exiting in an easterly direction. In a statement that would be echoed in the United States the following year, Swedish officials suggested that the number of reports of ghost rockets was probably exaggerated because of the publicity, but when those were eliminated from the mix there was still a solid core of reports.

Probably the most important event of August 1946 was the arrival of Lieutenant General James H. Doolittle, the man who had planned and led the first bombing raid on Tokyo just months after the beginning of the Second World War. Doolittle had been briefed by U.S. intelligence about the events in Sweden. It had been announced that Doolittle was in Sweden on business for Shell Oil Company, but documents released in later years have revealed the truth. American military officers were interested in the ghost rockets, and Doolittle had been dispatched to learn what he could about them.

During the last week in August, Norway joined the ranks of the countries reporting the ghost rockets. The *London Daily Telegraph* reported, "Two of the projectiles have been seen over Oslo during the past week, and two are reliably reported to have landed in Lake Mjoesen, north of the city. Unfortunately the lake is too deep for the authorities to be able to hold out any hope of dredging the pieces. . . . The Norwegian General Staff issued a statement to the press, asking it not to make any mention of the appearance of the rockets over Norwegian territory but to pass on all reports to the Intelligence Department of the High Command."

In September more of Europe became involved. In Greece, according to newspaper reports, "Acting Foreign Minister Stephanos Stephanopolos supported a statement in London by Premier Constantin Tsaldaris that flying rockets had been seen over Greece."

On September 17, north Africa entered the arena. From Morocco came the report that "a flying projectile with a 'tail of flame' was seen last night over the town of Fez; it was reported at Casablanca today."

Over the next few days, reports would be made in Germany, Portugal, Belgium, and Holland. The reports would detail points of light, elongated rockets, and even disc-shaped craft. The sightings included everything that would become a staple of the UFO waves

Another typical ghost rocket

that were to follow, including an investigation by governmental officials and attempts to suppress the information reported by the newspapers.

On October 10, 1946, the Swedish Defense Staff released its findings concerning the ghost rockets. They concluded that the majority of the sightings were of conventional objects seen in unusual circumstances. The staff also reported that "in some cases, clear, unambiguous observations have been made which cannot be explained as natural phenomena, Swedish aircraft, or imagination on the part of the observer. Echo, radar, and other equipment registered readings, but gave no clue as to the nature of the objects." The few samples of physical evidence that had been recovered by the military authorities were identified as routine and provided no clues to the identity of the ghost rockets.

If we are to believe what has been written before, the Swedish government actively suppressed the tales of the ghost rockets. If people were reporting them for the publicity they would receive, then the government was doing what it could to stop the wave. And it must be noted that the suppression of the reports did not stop people from making observations of the objects.

Of course, American officers would make the same claim about similar waves in this country. But it seems that few of those reporting sightings had visions of their names in the newspapers. While the reports might have been of astronomical phenomena, unusual weather events, or conventional aircraft seen under less than perfect conditions, few of those making the reports sought any sort of publicity. In fact, in many cases, the people in this country made their reports only when promised their names wouldn't appear in the newspapers.

With the Swedish and Scandinavian ghost rockets, we can trace the reports back beyond 1946 and the Second World War and into the 1930s, just as we can with the reports of the flying saucers that began in this country in 1947. In 1939, for example, a ghost-rocket report surfaced, for the first time, in a Swedish newspaper. Over the next few years, there would be other reports, but some of those seem to be traceable to the actual German experiments on the V-1s and V-2s.

However, the phenomenon seems to have originated in November and December 1933, near the border with Norway and just below the Arctic Circle. The sightings there were not of rockets, but of lights that seemed to fly or float along the valleys near the border. The only known aircraft in the area at the time could be accounted for, but there was the possibility that some sort of smuggling operation was taking place.

As the number of reports grew, and in some respect because of newspaper reports, the various Scandinavian governments tried to find explanations. The Norwegians, Swedes, and Finns all shared their information and cooperated with one another to either force down or capture one of these "mysterious airplanes."

Studies made by the various governments came to the same conclusions that modern authorities do when confronted with UFO sightings. Many are from reliable observers, but many, if not most, are from people unfamiliar with the sky around them. In a Swedish study of 487 cases reported during the winter of 1933–

1934, they wrote off nearly a quarter of them as "unbelievable." But more important, there was a solid core of reliable reports that couldn't be explained in a conventional sense.

Just over a decade later, the Swedish Defense Staff would be investigating nearly a thousand ghost-rocket reports. What separates the ghost rockets from other UFO waves is that nearly one hundred impacts of the craft were reported, and thirty pieces of debris from those crashes were submitted to the military for analysis. Those analyses failed to reveal anything of extraterrestrial origin.

One of the most important and the most interesting of those crash stories, was the July 19 event mentioned earlier. It was just before noon when a farmer, Kurt Lindback, and others, heard a rumbling that drew their attention to the sky. Lindback spotted something that he thought at first was a conventional airplane. It was about seven feet long and had two stubby wings.

The light-gray object fell into the lake about a mile from where he was standing. There was a large splash, followed by a second one, suggesting some sort of explosion. The object disappeared under the water.

Another witness, who was standing on the shore, told of the horrible noise. He said that it was as if a bomb had detonated nearby. Of course, there was no one dropping bombs on Sweden a year after the war had ended. He had no explanation for what he had seen.

The Swedish military did investigate at length. The military cordoned the area and erected a raft over the location where the object fell. From the raft they could see that something had detonated underwater. The lake bed was disturbed, and the plant life had been thrown up and out onto the nearby shore.

Two engineers arrived shortly after the military. They searched for signs of radioactivity but found none. The military search was finally suspended when nothing physical was located.

Although the sightings and investigations have been well documented by not only Swedish authorities but by other Scandinavian officials, no solution or explanation for the ghost-rocket epidemic has been found. The introduction of radar and other monitoring equipment only proved that something physical was seen in some of the locations. And in 1946 although many in Sweden thought that the ghost rockets were from the Soviet Union, no credible evidence has ever been found to support that conclusion. Instead, the Swedish officials were left with no reliable answer. All they knew was that something that resembled a modern cruise missile was being fired at and seen over much of Europe, but there was no one in 1946 who had the technology to produce such a weapon.

The stylized object that Kenneth Arnold reported

With the end of 1946 and the disappearance of ghost-rocket reports in Scandinavia, the sighting activity finally moved to the United States. There were a number of sightings of strange things in the sky but none achieved any sort of national attention until Kenneth Arnold, a Boise, Idaho, businessman, claimed to have seen nine crescent-shaped objects in formation flying near Mt. Rainier, Washington, on June 24, 1947. When reporters questioned him, he said that the objects moved with a motion like that of a saucer skipping across the water. The term "flying saucer" was born.

That wasn't the only such report. Within days of Arnold's sighting, thousands were claiming to have seen the flying saucers. Newspaper headlines from around the country screamed that the "Saucers" or the "Flying Disks," as they were often called, had been seen everywhere. None of these reports was more important than that from Roswell, New Mexico, in which military officers claimed to have "captured" a flying saucer with the help of a local rancher.

Within hours of that announcement, Army Air Forces was claiming that a mistake had been made and that the flying saucer was nothing more spectacular than a stray weather balloon and its rather common radar reflector. There are those today who suggest that this was the crash of an alien spacecraft and that a cover-up of the truth about the flying saucers began there in New Mexico.

The next day, newspapers around the country carried a story that the army and the navy had begun a campaign to stop the stories of flying saucers "whizzing" through the atmosphere. To many, the timing was suspect. One day they announce they had captured a flying saucer, and the very next, they begin to suppress the information.

It was becoming clear to those at the top of the military power structure that something had to be done. Yes, they had begun a campaign to stop the flow of information in the civilian world, but that didn't mean the military and the Pentagon weren't interested in learning all they could about the flying saucers. Although they had been investigating sightings during the summer of 1947 and had been involved in the analysis of the reports they'd gathered, something official was brewing. As the summer drew to a close, military officers realized that some sort of project designed to gather data was needed.

The first document to relate to the beginnings of that endeavor, eventually code named "Blue Book" was written and signed on September 23, 1947. Lieutenant General Nathan F. Twining, then the commander of the Air Material Command, suggested to the commanding general of the Army Air Forces, General Carl Spaatz, through a letter to Brigadier General George Schguler, that "the phenomenon reported is something real and not visionary or fictitious." Twining recommended that "Headquarters, Army Air Forces issue a directive assigning a priority, security classification and Code Name for a detailed study of the matter."

On December 30, 1947, Major General L. C. Craigie approved the recommendation. Project Sign, the forerunner of Blue Book, was created as a 2A classified project, the highest being 1A. Although the general public was aware that a project had been created, they knew it as "Project Saucer," rather than under the real name of Project Sign.

Although the project did not exist in the summer of 1947, its files did include cases from that period. The first recorded flying-saucer sighting in the official records is from early June and from Hamburg, New York. The case is missing, but according to the index, created during the project, it was a misidentification of an aircraft. The next case was from Seattle, Washington, and is also missing. Its information is listed as "Insufficient Data."

According to the documentation available today, Project Sign began its work on January 22, 1948 and conducted its first major investigation into the crash of a Kentucky National Guard aircraft flown by Captain Thomas Mantell. He had been chasing an unknown object near the Godman Army Air Field when the accident took place.

Air force investigation into the tragedy would take place on two levels. The first level concerned examining the event as an aircraft accident. The accident investigation board ruled that Mantell, in an F-51, had flown above twenty thousand feet without proper oxygen equipment in the aircraft. He had blacked out due to oxygen starvation, and the plane had gone into a power dive and crashed, killing Mantell. Almost no one questioned that, with the exception of some UFO believers who claimed Mantell had been shot down by the object he was chasing.

On the second level, the air force investigated the crash as a UFO incident because Mantell, at the time of his death, had been chasing a reported flying saucer. He had been diverted from a routine flight into an intercept mission when a number of people on the ground at Fort Knox, Kentucky, spotted a huge UFO over the airfield. Mantell and his wingmen set off to investigate.

Because of the altitude and fuel problems, one by one, the wingmen broke off the intercept. But Mantell could still see the UFO, believed that it was flying above him at about half his speed. He was determined to catch it and identify it.

Air force investigators ascertained that Venus was visible in the sky on the day Mantell died and suggested he had been attempting to intercept the planet. When that theory proved inadequate, they suggested he had chased a weather balloon. When that proved inadequate, they suggested a combination of Venus and a balloon, and finally Venus and two balloons.

The true nature of the UFO that Mantell chased was discovered several years later when the navy's classified cosmic-ray balloons, known as Skyhook, were revealed. Given the descriptions, the timing, and the situation, it is all but certain that Mantell had chased one of the Skyhook balloons.

By the summer of 1948, dozens of seemingly inexplicable cases had been reported to the Air Technical Intelligence Center at Wright Field in Dayton, Ohio. It wasn't until a DC-3 was "buzzed" by a flying saucer in July 1948 that the situation began to crystal-

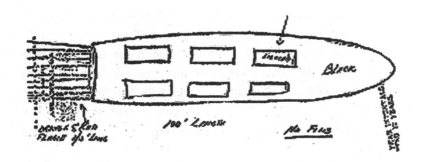

The flaming craft reported by Chiles and Whitted

lize. On July 24, a rocket-shaped object with two rows of square windows and flames shooting from the rear flashed past an aircraft piloted by Captain Clarence S. Chiles and John B. Whitted. At least one of the passengers also saw the object. An hour earlier, a ground maintenance crewman at Robins Air Force Base in Georgia claimed to have seen the same object. The sightings made a solid case of multiple witnesses who described, more or less, the same object. They also suggested that the flying saucers were not something of earthly origin.

Some of the officers at the official investigation, Project Sign, convinced they now had proof that the flying saucers were extraterrestrial, put together what they called "an Estimate of the Situation." They concluded that flying saucers came from other planets, wrote a report containing the best evidence they had, and shipped it up to General Hoyt S. Vandenberg, then air force chief of staff. According to Captain Edward Ruppelt, who eventually headed up Project Blue Book, Vandenberg wasn't impressed with the evidence. He rejected the report as not containing sufficient documentation to prove the case and ordered it to be declassified and then destroyed.

Dr. Michael Swords, who had an opportunity to review drafts of Ruppelt's original manuscript for *The Report on Unidentified Flying Objects,* made some interesting observations about the tale of the

Estimate of the Situation as told by Ruppelt. In an article originally published by the J. Allen Hynek Center for UFO Studies in their *International UFO Reporter*, Swords outlined the deletions that Ruppelt had made between the original version of his book and the version that was eventually published. To illustrate what had been left out of the book as published, Swords used italics to show the deleted material. Ruppelt's text as published was printed in roman typeface.

In intelligence, if you have something to say about some vital problem you write a report that is known as an "Estimate of the Situation." A few days after the DC-3 was buzzed, the people at ATIC decided that the time had arrived to make an Estimate of the Situation. The situation was UFO's; the estimate was that they were interplanetary!

It was a rather thick document with a black cover and it was printed on legal-sized paper. Stamped across the front were the words TOP SECRET.

It contained the Air Force's analysis of many of the incidents I have told you about plus many similar ones. All of them had come from scientists, pilots, and other equally credible observers, and each one was an unknown.

It concluded that UFO's were interplanetary. As documented proof, many unexplained sightings were quoted. The original UFO sighting by Kenneth Arnold; the series of sightings from the secret Air Force Test Center, MUROC AFB; the F-51 pilot's observation of a formation of spheres near Lake Mead; the report of an F-80 pilot who saw two round objects diving toward the ground near the Grand Canyon; and a report by the pilot of an Idaho National Guard T-6 trainer, who saw a violently maneuvering black object.

As further documentation, the report quoted an interview with an Air Force major from Rapid City AFB (now Ellsworth AFB) who saw twelve UFO's flying a tight diamond formation. When he first saw them they were high but soon they went into a fantastically high speed dive, leveled out, made a perfect formation turn, and climbed at a 30 to 40 degree angle, accelerating all the time. The UFO's were oval-shaped and brilliant yellowish-white.

Also included was one of the reports from the AEC's Los Alamos Laboratory. The incident occurred at 9:40 A.M. on September 23, 1948. A group of people were waiting for an airplane at the landing strip in Los Alamos when one of them noticed something glint in the sun. It was a flat, circular object, high in the northern sky. The appearance and relative size was the same as a dime held edgewise and slightly tipped, about 50 feet away.

The document pointed out that the reports hadn't actually started with the Arnold incident. Belated reports from a weather observer in Richmond, Virginia, who observed a "silver disk" through his theodolite telescope; an F-47 pilot and three pilots in his formation who saw a "silvery flying wing," and the English "ghost airplanes" that had been picked up on radar early in 1947 proved the point. Although reports on them were not received until after the Arnold sighting, these incidents had all taken place earlier.

When the estimate was completed, typed, and approved, it started up through channels to higher-command echelons. It drew considerable comment but no one stopped it on its way up.

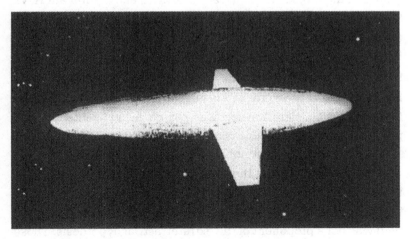

An example of what the English called a "ghost airplane"

General Vandenberg eventually received the estimate at the Pentagon but was apparently less than impressed with the evidence. According to Ruppelt, at that point, "it was batted back down. The general wouldn't buy interplanetary vehicles. The report lacked proof."

A group of military officers and civilian technical intelligence engineers was then called to the Pentagon to defend the estimate. According to the work done by Michael Swords, civilian group members were likely Lawrence H. Truettner, A. B. Deyarmond, and Alfred Loedding. Swords noted, parenthetically, that "Truettner and Deyarmond were the authors of the Project Sign report that contained many of these same cases and sympathies; Loedding was a frequent Pentagon liaison in 1947 and considered himself the 'civilian project leader' of Sign."

The military participants were probably the official project officer, Captain Robert Sneider, as well as Colonel Howard McCoy or Colonel William Clingerman, who would have had to sign off on the estimate.

Swords noted that the defense was unsuccessful, but not long after the visit to the Pentagon, everyone named was reassigned. Swords wrote, "So great was the carnage that only the lowest grades in the project, civilian George Towles and Lieutenant H. W. Smith, were left to write the 1949 Project Grudge document about the same cases."

Swords pointed out that Donald Keyhoe had mentioned the existence of the estimate a number of times and was told that it was a myth. According to Swords: "The famous Armstrong Circle Theatre fiasco of 1958, where Keyhoe was cut off in mid-sentence, was partly due to the fact that he was about to mention this document."

After Vandenberg "batted" the report back down, after the staff was reduced, and after the fire went out of the investigation, Project Sign limped along. It was clear to everyone inside the military, particularly those who worked around ATIC, that Vandenberg was not a proponent of the extraterrestrial hypothesis. Those who supported the idea risked the wrath of the number one man in the air force. They had just had a practical demonstration of that. If an officer was not smart enough to pick up the clues from what had just happened, then that officer's career could be severely limited.

Project Blue Book files show this to be the case. When Sign evolved into Project Grudge and then into Blue Book, a final report about Sign was written. Those inside Sign originally believed that UFOs were extraterrestrial until Vandenberg said he didn't find their reasoning adequate. Then, those who were left inside Sign decided that other answers must be the correct ones. UFOs, flying saucers, were not extraterrestrial.

A report entitled "The Findings of Project Sign" was eventually written. It outlined the motivation behind Project Sign, who the players were, and then presented the results of the research. In the "Summary," it was noted that the Sign data was "derived from reports of 243 domestic and thirty (30) foreign incidents. Data from these incidents is being summarized, reproduced and distributed to agencies and individuals cooperating in the analysis and evaluation. . . . The data obtained in reports received are studied in relation to many factors such as guided missile research activity, weather and other atmospheric sounding balloon launchings, commercial and military aircraft flights, flights of migratory birds, and other considerations, to determine possible explanations for sightings."

The authors of "The Findings of Project Sign" wanted to make the situation quite clear to those who would be reading it. They wrote, "Based on the possibility that the objects are really unidentified and unconventional types of aircraft, a technical analysis is made of some of the reports to determine the aerodynamic, propulsion, and control features that would be required for the objects to perform as described in the reports. The objects sighted have been grouped into four classifications according to configuration:

1. Flying disks, i.e., very low aspect ration aircraft.
2. Torpedo or cigar shaped bodies with no wings or fins visible in flight.

3. Spherical or balloon-shaped objects.
4. Balls of light.

The authors reported that "approximately twenty percent of the incidents have been identified as conventional aerial objects to the satisfaction of personnel assigned to Project 'Sign' in this Command. It is expected that a study of the incidents in relation to weather and other atmospheric sounding balloons will provide solutions for an equivalent number. . . . Elimination of incidents with reasonably satisfactory explanations will clarify the problem presented by a project of this nature.

"The possibility that some of the incidents may represent technical developments far in advance of knowledge available to engineers and scientists of this country has been considered. No facts are available to personnel at this Command that will permit an objective assessment of this possibility. All information so far presented on the possible existence of space ships from another planet or of aircraft propelled by an advanced type of atomic power plant have been largely conjecture."

The "Findings" authors provided a number of recommendations, writing, "Future activity on this project should be carried on at the minimum level necessary to record, summarize, and evaluate the data received on future reports and to complete the specialized investigations now in progress." They then add a phrase that too many UFO researchers have overlooked in the past. They write, "When and if a sufficient number of incidents are solved to indicate that these sightings do not represent a threat to the security of the nation, the assignment of special project status to the activity could be terminated."

This is a theme that would be repeated in one official UFO investigation after another. They would mention this aspect again and again. Each of the investigations, from Sign forward, had national security as its main concern. If national security wasn't threatened, then the question of reality became unimportant. And, as time passed, it became more likely to all those military investigators that no threat to the nation was posed.

The authors also wrote, "Reporting agencies should be impressed with the necessity for getting more factual evidence on sightings such as photographs, physical evidence, radar sightings, and data on size and shape."

The conclusions of the report were interesting: "No definite and conclusive evidence is yet available that would prove or disprove the existence of these unidentified objects as real aircraft of unknown and unconventional configuration. It is unlikely that positive proof of their existence will be obtained without examination of the remains of crashed objects. Proof of the nonexistence is equally impossible to obtain unless a reasonable and convincing explanation is determined for each incident."

The authors then wrote, "Many sightings by qualified and apparently reliable witnesses have been reported. However, each incident has unsatisfactory features, such as shortness of time under observation, distance from observer, vagueness of description or photographs, inconsistencies between individual observers, and lack of descriptive data, that prevent conclusions being drawn."

The reason for the recommendation for a continuation of the project had nothing to do with research into the UFO phenomenon. The authors wrote, "Evaluation of reports of unidentified objects is a necessary activity of military intelligence agencies. Such sightings are inevitable, and under wartime conditions rapid and convincing solutions of such occurrences are necessary to maintain morale of military and civilian personnel. In this respect, it is considered that the establishment of procedures and training of personnel is in itself worth the effort expended on this project."

About a year earlier, the personnel assigned to Sign had concluded that flying saucers were extraterrestrial. Now, using the same cases and the same evidence, those who survived at ATIC were claiming that there was nothing to the UFO phenomenon. More important, they were saying there was no threat to national security but that the project should be continued for training purposes.

There is another document, dated December 10, 1948, that was classified as top secret and detailed an earlier examination of UFO sightings. Air Intelligence Report No. 100-203-79, entitled "Analysis of Flying Object Incidents in the U.S.," apparently was a joint effort between the Directorate of Intelligence of the air force and the Office of Naval Intelligence. It is so sensitive that it contains a warning that states, "This document contains information affecting the national defense of the United States within the meaning of the Espionage Act, 50 U.S.C., 31 and 32, as amended. Its transmission or the revelation of its contents in any manner to an unauthorized person is prohibited by law. Reproduction of the intelligence in this publication, under the provisions of Army Regulation 380-5, is authorized for United States military agencies provided the source is indicated."

The document, then, was highly classified. It was a report created to brief high-ranking military officers on the unidentified flying object situation. It would seem that the officers creating the document would have access to all the classified information needed to accurately assess the sighting data.

The problem, according to the document was "TO EXAMINE patterns of tactics of 'Flying Saucers' (hereinafter referred to as flying objects) and to develop conclusions as to the possibility of existence."

Under discussion was, the report said, "THE POSSIBILITY that reported observations of flying objects over the U.S. were influenced by previous sightings of unidentified phenomena in Europe, particularly over Scandinavia in 1946, and that the observers reporting such incidents may have been interested in obtaining personal publicity have been considered as possible explanations. However, these possibilities seem to be improbable when certain selected reports such as the one from the U.S. Weather Bureau at Richmond are examined. During the observations of weather balloons at the Richmond Bureau, one well trained observer has sighted strange metallic disks on three occasions and another observer has sighted a similar object on one occasion. The last observation of unidentified objects was in April, 1947. On all four occasions the weather balloon and the unidentified objects were in view through a theodolite. These observations at the Richmond Bureau occurred several months before publicity on the flying saucers appeared in U.S. newspapers."

The report included an interesting paragraph about the origins of the objects. It said, "THE ORIGIN of the devices is not ascertainable. There are two reasonable possibilities: (1) The objects are domestic devices, and if so, their identification or origin can be established by a survey of the launchings of airborne devices . . . (2) Objects are foreign, and if so, it would seem most logical to consider that they are from a Soviet source. . . ."

The conclusions, at the bottom of page two, were marked top secret: "SINCE the Air Force is responsible for control of the air in the defense of the U.S., it is imperative that all other agencies cooperate in confirming or denying the possibility that these objects are of domestic origin. Otherwise, if it is firmly indicated that there is no domestic explanation, the objects are a threat and warrant more active efforts of identification and interception."

And finally, the report said, "IT MUST be accepted that some type of flying objects have been observed, although their identification and origin are not discernible. In the interest of national defense it would be unwise to overlook the possibility that some of these objects are of foreign origin."

What we observe in this document, however, is not the all-knowing, access to every classified report that skeptics have suggested. Instead, we find the authors speculating that the flying objects might be a domestic project and their suggestion that any such project should be revealed to the air force because of its responsibility for air defense. In other words, the authors of the top secret report did not have complete access to everything. They admitted that there were areas they were not allowed to examine and there were things they weren't allowed to learn. It suggests there might be other areas in which the UFO information was concealed that was outside the scope of their authority, no matter how highly placed that authority, or how high their security clearances were.

All "Analysis" suggests is that two officers assigned the task of investigating UFOs from an "insiders' perspective" found areas where they were not allowed to search because their clearances didn't provide a high-enough authority. They freely admitted this, and their predicament demonstrates one of the problems with compartmentalization. What one group is doing is not to be communicated to another group who might be working on an aspect of the same problem because that might result in a compromise of highly classified information.

The document also demonstrates the split personality of UFO investigations at that time. And it demonstrates the secrecy with which some investigations were being conducted. For example, on December 16, 1948, just a year after its beginning, Project Sign became Project Grudge. The old name had been compromised; that is, it was known to too many people who shouldn't have known it. The solution for the air force was an announcement that Project Sign had completed its assignment and had been terminated.

That wasn't the situation, however. Project Sign was now Project Grudge. With the new name came a new attitude. All reports were investigated on the premise that they were simply misidentifications of natural phenomena or aircraft or they were outright hoaxes. Flying saucers didn't exist, so there could be no proof found that they existed.

Just a year later, on December 27, 1949, the air force announced that Project Grudge, the official investigation into the flying-saucer sightings, was being closed. The final report would be available to reporters as soon as it was completed.

That final report contained the case studies of 237 of the best accounts. A number of experts, including Dr. J. Allen Hynek, were able to explain some of the sightings as astronomical phenomena. Captain A. C. Trakowski, of the air force's Cambridge facility, reviewed the balloon-flight records to determine if any of the sightings could be explained by appearances of various types of balloons being used at the time. Trakowski's review was an in-depth study.

And, when all was said and done, the Grudge report had explained all but 23 percent of the sightings. The Psychology Branch of the air force's Aeromedical Laboratory attempted to eliminate that 23 percent. The officials there concluded that "there are sufficient psychological explanations for the reports of unidentified flying objects to provide for plausible explanations for reports not otherwise explainable. . . ."

The point that seems to have gotten lost, however, is that nearly a quarter of the sightings reported to the air force did not have mundane explanations. So, when they failed to find a solid explanation, the air force invented the psychological category. As those at the Aeromedical Laboratory suggested, some people just had spots in front of their eyes for any number of reasons.

But even though the air force had announced that it had closed Project Grudge, such was not the case. Grudge continued to function at a low level with a single investigator, Lieutenant Jerry Cummings. When a series of spectacular sightings were made at the army's Signal Corps Center at Fort Monmouth, New Jersey, Cummings, and Lieutenant Colonel N. R. Rosengarten, who was the chief of ATIC's Aircraft and Missiles Branch, were sent to investigate. They then personally briefed the chief of air force intelligence, Major General C. P. Cabell.

The meeting didn't go well because Cabell, other military officers, and representatives of Republic Aircraft, complained about the quality of the work being done by Grudge. There was a perceived threat to national security, though no one was sure exactly what that threat might be. Cummings and Rosengarten were ordered back to Wright-Patterson Air Force Base with orders to reorganize and revitalize the UFO project.

Cummings didn't have much of a chance to do anything. He

was discharged from the air force shortly after the meeting. Rosengarten then asked Ed Ruppelt, an intelligence officer at ATIC, to reorganize Grudge.

Ruppelt believed his first task was to reorganize the files, coordinate the new reports being made, and cross-reference all of them. He found that many of the earlier cases were now missing from the files. He also put together a staff who had no firm beliefs for or against the idea that flying saucers were real and extraterrestrial. And he subscribed to a clipping service so that he would be able to learn of new sightings that were not reported to the military authorities for a variety of reasons. He hoped to gain some insights into the UFO problem by gathering as much data as possible about the phenomenon.

In 1950 and 1951 combined, there were 379 sightings reported to the air force UFO project. Of those, all but forty-nine were explained as mundane. Other sightings, from 1947 to 1949 were periodically reviewed, and new solutions were attached to the old cases as new information or additional facts were gathered.

In March 1952, Grudge had its status upgraded. Now it was the Aerial Phenomena Group and the code name had been changed once again. It was now known as Project Blue Book.

But just as Ruppelt was getting things organized, the situation changed. Instead of getting two or three UFO reports a week, they began to come in two or three a day. In his book, Ruppelt wrote that the clippings that had been coming in a thick envelope once a month began to arrive in boxes.

July 1952 would be the big month. On two consecutive weekends, UFOs were spotted over the Washington, D.C., area. The sightings began late in the evening on July 19 when two radars at the Air Routing and Traffic Control Center (ARTC) picked up eight unidentified targets near Andrews Air Force Base. According to reports made by the controllers, these objects were not airplanes because they moved too fast. One object, according to the calculations made by the controllers, was tracked at seven thousand miles an hour.

About twenty minutes later, just after midnight on July 20, the tower radars at Washington's National Airport tracked five objects. What this meant was that three radars at three different locations had solid targets that were not identified as aircraft.

One of the controllers at the ARTC called for a senior controller, Harry C. Barnes, who in turn called the National Airport control tower. The airport had unidentified targets on their scopes, as had the controllers at Andrews Air Force Base. They had already eliminated a mechanical malfunction as the cause, but with the objects on other scopes in other locations, there was no longer any question of their reality. The performance of the blips ruled out airplanes. All the witnesses, including Barnes, were sure they were looking at solid objects based on their years of experience as air traffic controllers and with radar. Weather-related phenomena wouldn't produce the same effect on all the radars at widely scattered locations. In fact, if weather was the explanation, the targets would have varied from scope to scope.

Just after midnight on July 20, Airman Second Class Bill Goodman called the Andrews Air Force Base control tower to tell them he was watching a bright orange light about the size of a softball that was gaining and losing altitude as it zipped through the sky. He was reporting a visual observation of a UFO at the same time that radar operators were watching the blips on their scopes.

During this time, Goodman talked to Airman First Class William B. Brady, who was in the Andrews tower. Goodman told Brady that the object was to the immediate south. Brady saw a ball of orange fire. There were discrepancies between the physical descriptions given by Goodman and Brady, but the differences were relatively small. It can be argued that the discrepancies were the result of the points of view of the two observers.

About two in the morning on July 20, the radar officer at Andrews Approach Control, Captain Harold C. Way, learned that the ARTC had a target east of Andrews. He went outside and saw a strange light which he didn't believe to be a star. Later, however, he went back out, and this time decided that he was, in fact, looking at a star.

Bolling Air Force Base became involved briefly about the time Way went outside. The tower operator there said that he saw a "roundish" object drifting low in the sky to the southeast of Bolling. There were no radar confirmations of the sighting, and that was the last of the reports from that base.

The ARTC again told the controllers at Andrews that they still had the targets on their scopes. There is conflicting data about what was detected because some of the reports suggested that the Andrews radar showed nothing, while other reports claim they did. Now Joe DeBoves, and two others in the Andrews tower, Monte Banning and John P. Izzo, Jr., swept the sky with binoculars but could see no lights other than the stars.

The sightings lasted through the night, and during that time, the crews of several airliners saw the lights right where the various radars showed them to be. Tower operators also saw them, and a jet fighter was brought in for attempted intercepts. Associated Press stories written hours after the sightings claimed that no intercepts had been attempted that night but those stories were inaccurate. Documents in the Project Blue Book files, as well as eyewitnesses, confirm the attempted intercepts.

Typical of the sightings were those made by Captain Casey Pierman on Capital Airlines Flight 807. He was on a flight between Washington, D.C., and Martinsburg, West Virginia, at 1:15 A.M. on July 20, when he and the rest of the crew saw seven objects flash across the sky. Pierman said, "They were like falling stars without trails."

Capital Airline officials said that National Airport radar picked up the objects and asked Pierman to keep an eye on them. Shortly after takeoff, Pierman radioed that he had the objects in sight. He was flying at 180 to 200 miles per hour and reported the objects were traveling at tremendous speed. Official air force records confirm this report.

Another Capital Airlines pilot, Captain Howard Dermott, on Capital Flight 610, reported a single light followed him from Herndon, Virginia, to within four miles of National Airport. Both the ARTC and the National tower confirmed that an unidentified target followed the aircraft to within four miles of landing. At about the same time, an air force radar at Andrews was tracking eight additional unknown objects as they flew over the Washington area.

One of the most persuasive sightings came early in the morning when one of the ARTC controllers called the Andrews control tower to tell them that there was a target south of the tower, over the Andrews radio range station. The tower operators looked to the south where a "huge fiery-orange sphere" was hovering. This again was later explained by the air force as a star.

Just before daylight, about four in the morning, after repeated requests from the ARTC to the Pentagon, an F-94 interceptor arrived on the scene, but it was too little too late. All the targets were gone. Although the flight crew made a short search of the local area, they found nothing unusual and returned to their base quickly.

During that night, apparently the three radar facilities only once reported a target that was seen by all three facilities. There were, however, a number of times when the ARTC radar and the Washington National tower radars had simultaneous contacts. It also seems that the radars were displaying the same targets that were seen by the crews of the Capital Airlines flights. What it boils down to is that multiple radars and multiple eyewitnesses were showing and seeing objects in the sky over Washington, D.C.

Air force intelligence, including ATIC and the officers assigned to the UFO project, had no idea that these sightings had taken place. They learned of the Saturday night–Sunday morning UFO show when the information was published in several newspapers on Monday. Ruppelt, on business in Washington and unaware of the sightings, reported "I got off an airliner from Dayton and I bought a newspaper in the lobby of Washington National Airport Terminal Building. I called Major Dewey Fournet, but all he knew was what he read in the papers."

Ruppelt wanted to stay in Washington to investigate the case but the bureaucracy got in the way. Ruppelt's orders didn't allow for an overnight stay. He tried to get them amended, but failed. He was warned that if he remained in Washington, even if on official business, he would be considered as absent without leave. Ruppelt had no choice but to return to Ohio without talking to anyone about the sightings.

A week later, almost to the minute, with the same crew on duty at National Airport, the UFOs returned. About ten-thirty P.M., the radar operators spotted several slow-moving targets. This time the controllers carefully marked each of the unidentifieds. When they were all marked, they called the Andrews AFB radar facility. The unidentified targets were on their scopes too.

An hour later, with targets being tracked continually, the controllers called for interceptors. Al Chop, the Pentagon spokesman for the UFO project in 1952, interviewed more than forty years later, said that he was in communication with the main basement command post at the Pentagon after he arrived at the airport. He requested that interceptors be sent. As a civilian, he could only make the request and then wait for the flag officer (general or admiral) in command at the Pentagon to make the official decision.

As happened the week before, there was a delay, but by midnight, two F-94s were on station over Washington. At that point, the reporters who had assembled to observe the situation were asked by Chop to leave the radar room at National Airport because classified radio and intercept procedures would be in operation.

Major Dewey Fournet, the Pentagon liaison between the UFO project in Dayton and the intelligence community in Washington was also at National Airport, having arrived after Chop. Fournet was accompanied by Lieutenant Holcomb of the Navy, an electronics specialist assigned to the air force Directorate of Intelligence. Holcomb's particular expertise was in radar operations.

72

With Chop, Fournet, and Holcomb watching, as well as the controllers at various facilities using various radars, the F-94s arrived. And the UFOs vanished from the scopes immediately. The jets were vectored to the last known position of the UFOs, but even though visibility was unrestricted in the area, the pilots could see nothing. The fighters made a systematic search of the vicinity, but since they could find nothing, they returned to their base.

Chop said, "The minute the first two interceptors appeared on our scope all our unknowns disappeared. It was like they just wiped them all off. All our other flights, all the known flights were still on the scope. . . . We watched these two planes leave. When they were out of our range, immediately we got our UFOs back."

Later, air force officers would learn that as the fighters appeared over Washington, people in the area of Langley Air Force Base, Virginia, spotted weird lights in the sky. An F-94 on a routine mission in the area was diverted to intercept the lights. The pilot saw one light and turned toward it, but it disappeared immediately "like somebody turning off a light bulb."

The pilot continued the intercept and did get a radar lock on the now unlighted and unseen target. That was broken by the object as it sped away. The fighter continued the pursuit, obtaining two more radar locks on the object, but each time the locks were broken.

The scene then shifted back to Washington National Airport. Again the Air Defense Command was alerted, and again fighters were sent. This time the pilots were able to see the objects, vectored toward them by the air traffic controllers. But the fighters couldn't close in on the lights. The pilots saw no external details, other than lights where the radar suggested that something should be seen.

After several minutes of failure to close in on a target, one of the objects was spotted loping along. Lieutenant William Patterson turned his plane, kicked in the afterburner, and tried to catch the object. It disappeared before Patterson could see much of anything.

Interviewed the next day, Patterson told reporters, "I tried to make contact with the bogies below one thousand feet, but they [the controllers] vectored us around. I saw several bright lights. I was at my maximum speed, but even then I had no closing speed. I ceased chasing them because I saw no chance of overtaking them. I was vectored into new objects. Later I chased a single bright light which I estimated about ten miles away. I lost visual contact with it. . . ."

Al Chop remembered this intercept, as did Dewey Fournet. Chop said, "The flight controllers had directed him [Patterson] to them [the unknowns]. We had a little cluster of them. Five or six of them and he suddenly reports that he sees some lights. . . . He said they are very brilliant blue-white lights. He was going to try to close in to get a better look . . . he flew into the area where they were clustered and he reported they were all around him."

Chop said that he himself, along with the others in the radar room, watched the intercept on the radar scope. What the pilot was telling them, they could see on the radar.

Patterson had to break off the intercept, though there were still lights in the sky and objects on the scope. According to Chop, the pilot radioed that he was running low on fuel. He turned so that he could head back to his base.

Chop said that the last of the objects disappeared from the scope about the time the sun came up. Ruppelt later quizzed Fournet about the activities that night. According to Ruppelt, Fournet and Holcomb, the radar expert, were convinced the targets were solid, metallic objects. Fournet told Ruppelt that there were weather-related targets on the scopes, but the controllers were ignoring them. Everyone was convinced that the UFO targets were real.

To explain the sightings over Washington, D.C., Major General John A. Samford, Chief of Air Intelligence, held a press conference the following Monday. Of that conference, Ruppelt writes,

General Samford made an honest attempt to straighten out the Washington National Sightings, but the cards were stacked

74

against him. He had to hedge on many answers to questions from the press because he didn't know the answers. This hedging gave the impression that he was trying to cover up something more than just the fact [that] his people fouled up in not fully investigating the sightings. Then he brought in Captain Roy James from ATIC to handle all the queries about radar. James didn't do any better because he'd just arrived in Washington that morning and didn't know very much more about the sightings than he'd read in the papers. Major Dewey Fournet and Lieutenant Holcomb, who had been at the airport during the sightings, were extremely conspicuous by their absence. . . .

As was the Pentagon spokesman on UFOs, Al Chop.

From that point, after the press conference, it seems that there was a semiofficial explanation for the Washington National sightings. One of the experts at the press conference, probably Roy James, suggested that a temperature inversion might have caused the unknown blips on the radar. UFO researchers have suggested that the temperature inversion theory simply doesn't work, especially when it is remembered that there were both radar and visual observations. But the air force officers were now happy. The news media had been given an explanation for the reports, and the sightings could now be ignored.

More important, Ruppelt, who did investigate, to some extent, the sightings, pointed out that air force personnel were pressured by their superiors to change their stories. Lights that had been inexplicable became stars seen through the haze hanging over the city. Skeptics suggested the radar returns were the result of the temperature inversion layers. It made no difference that the men on the scopes, and one of the military officers, were experts and could tell weather phenomena from solid targets. The sightings were explained, in the public arena, as temperature inversion. Curiously, the Blue Book files listed them as "unidentified."

Press and public interest increased dramatically when newspapers bannered the Washington National sightings at the end of July. During August the sightings throughout the country continued, as they did in September. By the end of 1952, the air force had added more than 1,500 sightings to the files. More important, more than three hundred of them were unidentified. The situation had become intolerable.

Part of the solution for investigating the sightings was the creation of the Robertson Panel. In September 1952, as the UFO reports were still flooding Project Blue Book, H. Marshal Chadwell, then Assistant Director of Scientific Intelligence, sent a memo to General Walter Bedell Smith, then Director of Central Intelligence (DCI) for the CIA. Chadwell wrote, "Recently an inquiry was conducted by the Office of Scientific Intelligence to determine whether there are national security implications in the problem of 'unidentified flying objects,' i.e. flying saucers; whether adequate study and research is currently being directed to this problem in its relation to such national security implications; and further investigation and research should be instituted, by whom, and under what aegis."

Chadwell continued, writing, "Public concern with the phenomena indicates that a fair proportion of our population is mentally conditioned to the acceptance of the incredible. In this fact lies the potential for the touching-off of mass hysteria. . . . In order to minimize risk of panic, a national policy should be established as to what should be told to the public regarding the phenomena."

It is clear from the tone of the document what Chadwell already believed. Tales of flying saucers are incredible. These things simply could not exist and therefore did not exist.

In December 1952, Chadwell decided to form a scientific advisory board. It was decided that Dr. H. P. Robertson, who had accompanied Chadwell to Wright-Patterson Air Force Base to review the UFO evidence, would chair the investigation.

Under the auspices of the CIA, the Robertson Panel convened on January 14, 1953. They reviewed the best UFO cases, including the films of UFOs that had been made by private citizens. They examined the best radar cases and the best photographic cases.

On Friday afternoon, with all the evidence finally presented, including briefings by Ruppelt and Hynek, Robertson began the task of writing a final report on the Panel's findings. By the next morning, in an age that had no copy machines, FAX machines, word processors, or computers, Robertson finished his draft of the official report. Not only that, Lloyd Berkner, one of the Panel members, had already read it, as had Chadwell, who had taken it to the Air Force Directorate of Intelligence to have it approved. Before the Panel assembled on Saturday morning, the report was, in essence, finished and approved. All done in a matter of hours after Robertson had written it.

One other thing must be understood to keep the Robertson Panel in perspective. Their first concern was to determine if UFOs posed a threat to national security. That was a question they could answer. They decided, based on the number of UFO reports made through official intelligence channels over the years, that UFOs did, after a fashion, pose a threat.

Ed Ruppelt mentioned this issue in his analysis of the Robertson Panel. Too many reports at the wrong time could mask a Soviet attack on the United States. Although hindsight shows us this supposed threat was of little importance, especially when the sorry state of Soviet missile research in 1952 is considered, it was a major concern to those men in the intelligence field in the early 1950s: A sudden flood of UFO reports, not unlike what had happened during the summer of 1952, could have created havoc in the message traffic so that critical messages of an imminent attack would have been hidden or lost.

With defense communication as a major concern, the Robertson Panel, who had seen nothing to suggest that UFOs were anything other than misidentifications, hoaxes, and weather and astronomical phenomena, needed to address this issue. That was the motivation behind some of the Panel's recommendations. These recommendations, then, were born of a need to clear the intelligence reporting channels and not of a need to answer the questions about the reality of the UFO phenomena.

In the Panel report, Robertson stated that "although evidence of any direct threat from these sightings was wholly lacking, related dangers might well exist resulting from: a. Misidentification of actual enemy artifacts by defense personnel. b. Overloading of emergency reporting channels with 'false' information ('noise to signal ratio' analogy—Berkner). c. Subjectivity of public to mass hysteria and greater vulnerability to possible enemy psychological warfare."

Robertson went on, writing, "Although not the concern of the CIA, the first two of these problems may seriously effect the Air Defense intelligence system, and should be studied by experts, possibly under ADC. If U.F.O.'s become discredited in a reaction to the 'flying saucer' scare, or if reporting channels are saturated with false and poorly documented reports, our capability of detecting hostile activity will be reduced. Dr. Page noted that more competent screening or filtering of reported sightings at or near the source is required, and that this can best be accomplished by an educational program."

Of all the suggestions in the Panel report, education is the issue that has caused the most trouble with interpretation. The Panel was suggesting that if people were more familiar with what was in the sky around them, if they were familiar with natural phenomena that were rare but spectacular, then many sighting reports could be eliminated. How many UFO sightings are explained by Venus, meteors, or bright stars that seem to hover for hours? In today's environment, with video cameras everywhere, how many times has Venus been taped and offered by witnesses as proof they saw something?

Under the subheading of "Educational Program," Robertson explained that "the Panel's concept of a broad educational program integrating efforts of all concerned agencies was that it should have two major aims: training and 'debunking.'"

Robertson continued, writing, "The training aim would result in proper recognition of unusually illuminated objects (e.g. balloons, aircraft reflections) as well as natural phenomena (meteors, fireballs, mirages, noctilucent clouds). Both visual and radar recognition are concerned. There would be many levels in such education. . . . This training should result in a marked reduction in reports caused by misidentified cases and resultant confusion."

The problem with the next paragraph of the Panel report came from the use again of the word *debunking*. Many have read something nefarious into it, but the use of the term and the tone of the paragraph suggest something that was, at the time, fairly innocuous, at least according to the Condon Committee established sixteen years later: "The 'debunking' aim would result in reduction in public interest in 'flying saucers' which today evokes a strong psychological reaction. This education could be accomplished by mass media such as television, motion pictures, and popular articles. Basis of such education would be actual case histories which had been puzzling at first but later explained. As in the case of conjuring tricks, there is much less stimulation if the 'secret' is known. Such a program should tend to reduce the current gullibility of the public and consequently their susceptibility to clever hostile propaganda. The Panel noted that the general absence of Russian propaganda based on a subject with so many obvious possibilities for exploitation might indicate a possible Russian official policy."

The Panel then discussed the planning of the educational program. Some believed the Panel's intent was to create a "disinformation" program designed to explain UFOs as mundane. The real reason behind the program, however, seems to have been to end sighting reports made by those who were unfamiliar with the sky. The educational program was suggested as a teaching tool.

The UFO information presented, according to those who were at some or all of the Panel's sessions, was "managed." They had a limited time to investigate and were unable to examine all aspects of the UFO field due to that constraint. It can be suggested that a careful management of the data supplied would provide a biased picture and that the conclusions drawn from that specific data could be accurate, but those conclusions would be skewed.

A careful study of the data supplied to the Robertson Panel does suggest that UFOs are little more than anecdotal gossip. The exceptions supplied to them are the UFO movies and the data from radar. However, without another piece of data, without some kind of physical evidence that would lead to the extraterrestrial hypothesis, no other conclusions could be drawn. The films were interesting, but there were alternative explanations, and while not as satisfactory in the long run, they were certainly no less valid. And radar cases are open to the interpretation of the radar operators. Their training, talent, and expertise are all important factors when considering that data.

It was at this same time, the beginning of 1953, that the investigative emphasis, which had dominated Blue Book for the eighteen months that had proceeded the Robertson Panel, began to erode. Although Project Blue Book would continue, its investigative responsibility was transferred to the 4602d Air Intelligence Service Squadron. Ruppelt suggests that the transfer was due to his demands that more investigators be found, but it seems that Blue Book was becoming too visible and too public.

Air Force Regulation 200-2 was in the planning stages with a version approved in August 1953. A year later, August 1954, the regulation went into effect, eliminating Blue Book from the investigative mix. Although the regulation required that ATIC be notified about the UFO investigations, there was nothing in it that required Blue Book to be informed. What we have is the classic situation where one agency has the responsibility for the majority of the UFO investigations and another has the authority. Blue Book had been effectively eliminated, though it still existed.

Ruppelt left the project, and it was handed off to a variety of other officers. At one low point in 1953, it was being run by an airman first class, a rather low enlisted grade. Ruppelt inherited it again for a couple of months, and then he was replaced. In March 1954 Captain Charles Hardin became the director.

In April 1956, Captain George T. Gregory, a man who didn't believe UFOs were anything other than misidentifications, led Blue Book into an almost rabid anti-UFO direction. The change in tone is evidenced in the limited investigations being conducted. During this time, sightings were to be identified, no matter how. The belief was that UFOs were not extraterrestrial spacecraft, and if they weren't, then another, mundane explanation should be available. The list of early sightings explained under this new concept is extraordinary. Other possibilities were left out of the case files so that it seemed, to those unfamiliar with the reports, that Blue Book's explanations were definite.

In December 1958, one of the officers assigned to the UFO project claimed that he found "certain deficiencies" that he felt "must be corrected." Specifically he referred to Air Force Regulation 200-2, "dated 5 February 1958, (revised on that date) which essentially stipulates the following . . . to explain or identify all UFO sightings."

After December 1958, there was an attempt to transfer Blue Book to some other air force agency, specifically, the Secretary of the Air Force, Office of Information (SAFOI).

On April 1, 1960, in a letter to Major General Dougher at the Pentagon, A. Francis Archer, a scientific advisor to Blue Book commented on a memo written by Colonel Evans, a ranking officer at ATIC. Archer said, "[I] have tried to get Bluebook out of ATIC for ten years . . . and do not agree that the loss of prestige to be a disadvantage."

In 1962 Lieutenant Colonel Robert Friend, who at one time headed Blue Book, wrote to his headquarters that the project should be handed over to a civilian agency that would word its report in such a way as to allow the air force to drop the study. At the same time, Edward Trapnell, an assistant to the secretary of the air force, when talking to Dr. Robert Calkins of the Brookings

Institution, said pretty much the same thing: Find a civilian committee to study the problem, then have them conclude it the way the air force wanted it to be resolved. One of the stipulations for this organization would be that it had to say some positive things about the air force's handling of the UFO investigations.

Other government officials suggested closing Blue Book but realized that the public would have to be "educated to accept the closing." By 1966 the air force managed to get Blue Book press releases made by SAFOI. Letters to the public no longer carried the prestigious ATIC or Foreign Technology Division letterhead but only the stamp of the Office of Information.

The major stumbling block to ending Project Blue Book was a new wave of sightings that were getting national attention. First, New Mexico police officer Lonnie Zamora reported an egg-shaped object on the ground near Socorro. He reported seeing two beings near it, and when it took off, it left landing-gear markings and burned vegetation. There was indirect physical evidence left by the object.

The public interest in UFOs began to rise. Network television paid attention, and several prestigious magazines began to treat the subject with a little respect. Air force explanations seemed tired, and even the most superficial investigations revealed flaws in their solutions. When Hynek, after hearing about the sightings in Michigan in 1966, said they might be the result of swamp gas, all credibility was lost.

Something had to be done because of the growing publicity. The air force was in a hole, and no one was listening. Someone decided that it was finally time for another independent study of the phenomena. The outgrowth of this was the Condon Committee, organized at the University of Colorado by Dr. Edward U. Condon and funded by more than half a million dollars of taxpayer money funneled through the air force.

Condon, the scientific director of the project and the man who received the air force grant, was a professor of physics and astrophysics, and a Fellow of the Joint Institute for Laboratory Astrophysics at the University of Colorado. As a career scientist, Condon had the sort of prestige the air force wanted.

As noted by the documentation that appeared after the declassification of the Project Blue Book files, and as noted here, the formation of the Condon Committee was part of an already existing plan. Find a university to study the problem (flying saucers) and then conclude it the way the Air Force wished.

Jacques Vallee, writing about the Condon Committee in *Dimensions,* said, "As early as 1967, members of the Condon Committee were privately approaching their scientific colleagues on other campuses, asking them how they would react if the committee's final report to the Air Force were to recommend closing down Project Blue Book." This comment tends to confirm that the real mission of Condon was not to study the UFO phenomena but to study ways to end air force involvement in it.

Dr. Michael Swords has spent the last several years studying the history of the Condon Committee and confirms the view that the air force used Condon for this purpose. But Condon was a willing participant in the deception. According to a letter discovered by Swords, written by Lieutenant Colonel Robert Hippler to Condon, the plan was laid out in no uncertain terms. Hippler told Condon that no one knew of any extraterrestrial visitation and therefore, there "has been no visitation."

Hippler also pointed out that Condon "must consider" the cost of the investigations of UFOs and to "determine if the taxpayer should support this" for the next ten years. If they failed to end the project in 1969, Hippler warned that it would be another decade before another independent study could be mounted that might end the air force UFO project.

Condon understood what Hippler was trying to tell him. Three days later in Corning, New York, Condon, in a lecture to a group of scientists, including those members of the Corning Section of the American Chemical Society and the Corning Glass Works Chapter of Sigma XI, told them, "It is my inclination right now to recommend that the government get out of this business. My attitude right now is that there is nothing in it. But I am not supposed to reach a conclusion for another year."

Robert Low responded to a letter from Hippler a day or so after Condon's Corning talk, telling him that they, the committee, were very happy that they now knew what they were supposed to do. Low wrote, "You indicate what you believe the Air Force wants of us, and I am very glad to have your opinion." Low pointed out that Hippler had answered the questions about the study "quite directly."

In 1969 the Condon Committee released their findings. As had all of those who had passed before them, the Condon Committee found that UFOs posed no threat to the security of the United States. Edward U. Condon in "Section I, Recommendations and Conclusions," wrote, "The history of the past 21 years has repeatedly led Air Force officers to the conclusion that none of the things seen, or thought to have been seen, which pass by the name UFO reports, constituted any hazard or threat to national security."

After stating that a UFO finding was "out of our [the Committee's] province" to study, and if they did find any such evidence, they would pass it on to the air force, Condon wrote, "We know of no reason to question the finding of the Air Force that the whole class of UFO reports so far considered does not pose a defense problem."

Also included in the "Recommendations," was Condon's statement that "it is our [the Committee's] impression that the defense function could be performed within the framework established for intelligence and surveillance operations without the continuance of a special unit such as Project Blue Book, but this is a question for defense specialists rather than research scientists."

That assessment seems to have taken care of most of the air force's requirements: Condon had confirmed that national security wasn't an issue, had said some positive things about the air force's handling of the UFO phenomena, and had recommended the end of Project Blue Book. He had done his job.

Finally, Condon wrote, "It has been contended that the subject has been shrouded in official secrecy. We conclude otherwise. We have no evidence of secrecy concerning UFO reports. What has been miscalled secrecy has been no more than an intelligent policy of delay in releasing data so that the public does not become confused by premature publication of incomplete studies or reports."

It is impossible to understand how Condon could write those words after being handed a stack of Blue Book files stamped top secret, which had been held by the air force for more than a decade. It is impossible to understand this, when, there was documentation that proves secrecy on the part of the air force. It was in 1969, before the official end of the Condon Committee, that Brigadier General C. H. Bolender wrote, "Moreover, reports of unidentified flying objects which could affect national security are made in accordance with JANAP 146 or Air Force Manual 55-11, and are not part of the Blue Book system."

In other words, documentation existed to support the claim there was secrecy. While a case can be made that the regulations and the secrecy are warranted by the circumstances, it can also be argued that the secrecy did exist, contrary to what Condon wrote.

The proof of secrecy demonstrates that the Condon Committee was not an unbiased scientific study of UFOs, but a carefully designed project that had a single objective: End public air force involvement in the UFO phenomena. After all, according to Hippler, should the taxpayers fund another ten years of UFO research?

The Condon report suggested there was no evidence of extraterrestrial visitation and that all UFO reports could be explained if sufficient data had been gathered in the beginning. This is exactly what Hippler wrote in his January 1967 letter to Condon. Yet, even when the Committee selected the sightings they would investigate, they failed to explain almost 30 percent of them. In one case (over Labrador, June 30, 1954), they wrote, "This unusual sighting should therefore be assigned to the category of some almost certainly natural phenomenon, which is so rare that it apparently has never been reported before or since."

But even with the holes in the study, even with the contradictory evidence, and even with the proof that something unusual was going on, Condon did what he was paid to do. He ended Project Blue Book. On December 17, 1969, the air force announced that it was terminating its study of flying saucers. The twenty-two-year-old study had come to a close.

The First UFO Account:

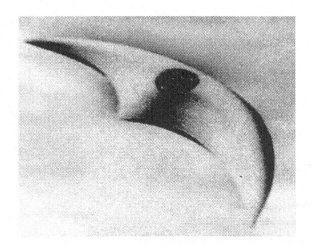

A highly stylized rendition of the Arnold crescent

Not long after the sighting, Arnold produced this drawing for military investigators.

Location: Mt. Rainier, Washington

Witnesses: Kenneth Arnold

Craft Type: Crescent-shaped

Dimensions: Thirty feet in diameter

Primary Color: Gray

Sound: None

Exhaust: None

Quality of Photos: Illustrations

Number of Photos: Illustrations

Type of Camera: N/A

Sources: Project Blue Book files; *Project Blue Book Exposed* by Kevin Randle; *The UFO Encyclopedia*, Volume 1, by Jerry Clark;

The Coming of the Saucers by Kenneth Arnold; *Watch the Skies* by Curtis Peebles; *The World of Flying Saucers* by Donald H. Menzel and Lyle G. Boyd; *The UFO Enigma* by Donald H. Menzel and Ernest H. Taves; *Scientific Study of Unidentified Flying Objects* edited by Daniel S. Gillmor

Relation to Other Sightings: Rhodes of July 7, 1947

Reliability: 6

Narrative: On June 24 the modern UFO era was ushered in when Kenneth Arnold, a Boise, Idaho, businessman saw nine objects flash across the sky near Mt. Rainer in Washington state. They were flying one behind the other, at about 9,500 feet at a speed estimated by Arnold to be more than 1,500 miles per hour.

Relating the tale later, Arnold told military investigators that "the air was so smooth that day that it was a real pleasure flying, and as most pilots do when the air is smooth and they are flying at a higher altitude, I trimmed out my airplane in the direction of Yakima, Washington, which was almost directly east of my position and simply sat in my plane observing the sky and the terrain."

His attention was drawn to the strange objects when the sunlight flashed off a metal surface. "It startled me, as I thought I was too close to some other aircraft. I looked every place in the sky and couldn't find where the reflection had come from until I looked to the left . . . where I observed a chain of nine peculiar looking aircraft."

The string of nine objects were flying in a formation that he estimated to be five miles long. They dodged in and out of the mountain peaks in a fluid motion that tilted them up and revealed their bottoms to him. He noted that they were quite far away.

Arnold had also seen a DC-4, which he estimated to be fifteen miles from him. He compared the objects to that aircraft, believing them to be smaller than the four-engine, propeller-driven airplane.

When he landed in Yakima, Washington, he told the assembled

reporters that the objects moved with a motion like that of saucers skipping across the water. The shape, however, according to drawings that Arnold completed for the army, showed objects that were heel-shaped. In later drawings, Arnold elaborated, showing objects that were crescent-shaped with a scalloped, trailing edge.

Hearing Arnold's description of the motion of the objects, reporter Bill Bequette coined the term "flying saucer." The term, then, didn't refer to the shape of the objects, but to the style of their movement.

Arnold's sighting didn't gain front-page status immediately. Stories about it appeared in newspapers a day or two after. It was, at that time, the story of an oddity. Arnold claimed later that he thought he had seen some sort of new jet aircraft.

Because this was the first of the flying-disc sightings to gain attention, it became important for military officers to determine what he had seen. They spent time and effort investigating it, and then wrote it off as mirages: That is, Arnold, because of the atmospheric conditions that afternoon, had seen a mirage in which the tops of the mountains seemed to be separated from the rest of the ground. It looked as if huge bits of land were hovering above the ground.

In a report prepared for the army air forces, Arnold expressed his displeasure at such suggestions. He wrote, "A number of news men and experts suggested that I might have been seeing reflections or even a mirage. This I know to be absolutely false, as I observed these objects not only through the glass of my airplane but turned my airplane sideways where I could open my window and observe them with a completely unobstructed view."

Arnold's assertion, of course, didn't satisfy those who believed that Arnold had made an error. Dr. J. Allen Hynek, the onetime consultant to Project Blue Book, studied the case for the military. It was Hynek's opinion that if Arnold's estimate of the distance was correct, then he had to have underestimated the size of the objects. If, on the other hand, he had overestimated the distance, then his timing of their flight was wrong. Hynek believed, according to the documents available, that the objects were closer than

I have received lots of requests from people who told me to make a lot of wild guesses. I have based what I have written here in this article on positive facts and as far as guessing what it was I observed, it is just as much a mystery to me as it is to the rest of the world.

My pilot's license is 333487. I fly a Callair airplane; it is a three-place single engine land ship that is designed and manufactured at Afton, Wyoming as an extremely high performance, high altitude airplane that was made for mountain work. The national certificate of my plane is 33355.

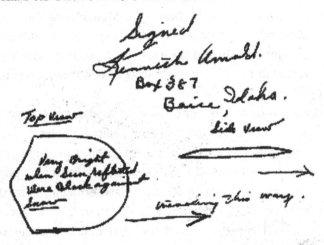

Page from the Project Blue Book report about the Arnold sighting showing his original drawing

Arnold thought. Hynek wrote, "In all probability, therefore, objects were much closer than thought and moving at definitely 'subsonic' speeds."

Others at Air Material Command (AMC), apparently impressed with Hynek's analysis, also wrote off the case. In their summary of the flying-saucer reports, that is, the Project Sign analysis, someone wrote, "AMC Opinion: The report cannot bear even superficial examination, therefore, must be disregarded. There are strong indications that this report and its attendant publicity is largely responsible for subsequent reports."

To those looking at the Arnold report in the late 1940s, it seemed to indicate that Arnold had misidentified some kind of known subsonic aircraft. But the question remains: What were they? The description of them fits nothing in the military inventory at the time, with the possible exception of the Northrop Flying Wing. It was a large, four-engine, propeller-driven aircraft that was not flying in the area of Arnold's sighting. And, there weren't nine Northrops available even if they had been flying.

There is another aspect of the case that needs to be clarified. In the air force file on the Arnold sighting, there are "galley proof" pages from a book written by Donald H. Menzel, the Harvard astronomer who believed that all UFO sightings were misidentifications or outright lies. In the book, Menzel proposes the mirage theory that the air force eventually accepted as the answer to the Arnold case.

But Menzel wasn't done. In his first book about flying saucers, Menzel suggested that Arnold had seen "billowing blasts of snow, ballooning up from the tops of ridges. . . . These rapidly shifting, tilting clouds of snow would reflect the sun like a mirror . . . and the rocking surfaces would make the chain sweep along something like a wave, with only a momentary reflection from crest to crest."

It is an interesting theory and one that makes sense, except that longtime residents say that in late June, what snow there is in the mountains is wet and heavy and wouldn't sweep around like the powdery stuff that falls in the winter. In other words, Menzel's explanation does not conform to the weather of the time, nor does it account for Arnold's description of the craft.

Menzel, apparently realizing the flaws in his theory, offered the possibility that a high layer of fog, haze, or dust just above or just below Arnold's altitude might account for the sighting. Menzel claimed that these layers could also reflect the sun in almost mirrorlike fashion.

Again the explanation fails, if only because Arnold saw movement and that would require some sort of turbulence at that altitude. Arnold had remarked about how stable the air was. A perfect day for flying with no real winds, or turbulence, and unlimited visibility.

Menzel, in his book with Lyle G. Boyd, *The World of Flying Saucers,* wrote that Arnold may have seen orographic clouds. These are huge, circular-shaped clouds that can form on the downwind side of mountains. But, as Menzel himself noted, they stand still and are not particularly reflective. In other words, Menzel, after suggesting the clouds, then eliminates them himself.

In the book *The UFO Enigma,* published after Menzel's death, and coauthored with Ernest H. Taves, Menzel suggested that Arnold may have been fooled by drops of water on the cockpit windows. He wrote, "I cannot, of course, say definitely that what Arnold saw were merely raindrops on the windows of his plane. He would doubtless insist that there was no rain at the altitude at which he was flying. But many queer things happen at different levels in the earth's atmosphere."

But remember what Arnold said about those who had suggested mirages: "I observed these objects not only through the glass of my airplane but turned my airplane sideways where I could open my window and observe them with a completely unobstructed view." If the objects had been water drops on the windows, they would have disappeared when he opened the window for his unobstructed view.

So, the real problem with Arnold's account seems to be that he may have underestimated the size of the objects or overestimated the distance to them. These are not fatal flaws. And they are no reason to throw out his report. Arnold was, after all, a pilot who had flown in the area before. He was familiar with what the terrain looked like. Instead of nit-picking Arnold's estimates of distance and size of the objects, or inventing multiple explanations that are contradicted by the facts, the military investigators should have been looking for corroboration of the case.

What is left is an interesting case that has no good explanation for the sighting it describes. The explanations posited are badly flawed or so outrageous as to be useless. This only means that Arnold saw something unconventional that, at the time, was not explained as aircraft, mirages, blowing snow, or raindrops on the cockpit canopy.

The First "Good" Photographs:

JULY 7, 1947

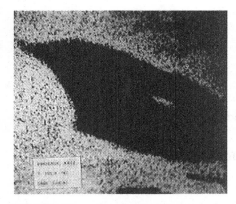

The Rhodes photographs seem to
show an object similar to the one
that crashed at Roswell.

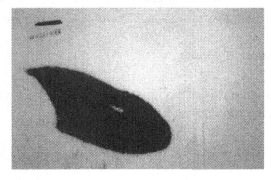

The second of the Rhodes photographs

Location: Phoenix, Arizona

Witnesses: William A. Rhodes

Craft Type: Crescent- or elliptical-shaped

Dimensions: Twenty to thirty feet in diameter

Primary Color: Gray

Craft Description: There seemed to be a "cockpit canopy in the center which extended toward the back and beneath the object. The cockpit did not protrude from the surface but was clearly visible with the naked eye." There were no propellers or landing gear.

Sound: Whoosh

Exhaust: None, though there seemed to be a wake of turbulent air that trailed the craft.

Quality of Photos: Sharp and clear

Number of Photos: 2

Type of Camera: Unavailable

Sources: Project Blue Book files; *Project Blue Book Exposed* by Kevin Randle; *The Arizona Republic* (July 1947)

Relation to Other Sightings: Kenneth Arnold, June 24, 1947; Fred Johnson, June 24, 1947; Roswell, New Mexico, July 4, 1947

Reliability: 7

Narrative: Less than two weeks after Kenneth Arnold's sighting was reported around the country, a self-employed scientist from Phoenix, Arizona, reported that he had taken what might be considered the first good photographs of one of the flying discs.

William A. Rhodes claimed he had been on the way to his workshop at the rear of his house when he heard a distinctive "whoosh" that he believed to be from a P-80 Shooting Star, a jet-powered aircraft. He grabbed his camera from a workshop bench and hurried to a small mound in the backyard. The object was circling in the east at about a thousand feet in the air.

Rhodes sighted along the side of his camera and took his first photograph. He advanced the film, and then hesitated, thinking that he would wait for the object to get closer. Then, worried that it would disappear without coming closer, snapped the second picture, finishing his roll of film.

Rhodes's story, along with his pictures, appeared in *The Arizona Republic*, a Phoenix newspaper. In the article, reporter Robert C. Hanika wrote, "Men long experienced in aircraft recognition studied both the print and the negative from which they [the photographs] were made, and declined to make a guess on what the flying object might be."

Hanika also wrote, "The marked interest Rhodes has for all air-craft has led most persons who have been in contact with other observers of the 'flying discs' to believe the photographs are the first authentic photographs of the missiles, since Rhodes easily can identify practically any aircraft."

Rhodes said that the object appeared to be elliptical in shape and have a diameter of twenty to thirty feet. It appeared to be at five thousand feet when first seen and was traveling, according to Rhodes, at four hundred to six hundred miles an hour. The craft was gray, and the color tended to blend with the overcast back-ground of the sky.

The object had, according to Rhodes in an interview for a confi-dential report in the Project Blue Book files, "what appeared to be a cockpit canopy in the center which extended toward the back and beneath the object. The 'cockpit' did not protrude from the surface but was clearly visible with the naked eye." There were no propellers or landing gear, but there did seem to be trails of turbu-lent air behind the trailing points of the object. There was specu-lation that there were jet engines of some kind located at those points. The craft moved silently, although Rhodes had said that a jetlike roar was what called his attention to it.

The news stories apparently alerted the military to Rhodes's sighting. Various investigations were launched. On July 14, 1947, in a memo for the record available in the Project Blue Book files, Lynn C. Aldrich, a special agent for the army's Counterintelligence Corps (CIC), wrote, "On 8 July 1947, this Agent obtained pictures of unidentifiable objects (Exhibits 1 and 2) from the managing ed-itor of the Arizona Republic newspaper. The pictures were taken by Mr. William A. Rhodes . . . [of] Phoenix, Arizona, at sunset, on 7 July 1947."

Then, on August 29, according to a "Memorandum for the Office in Charge," George Fugate, Jr., a special agent of the CIC and stationed at Forth Air Force Headquarters, interviewed Rhodes in person. Fugate was accompanied by Special Agent

Brower of the Phoenix FBI office. This interview is important because of some of the confusion about the location of the negatives and prints of the photographs that would arise later.

During the interview, Rhodes again told the sighting story, stating that he thought when he saw the object it might have been the navy's Flying Flapjack, which had been featured on the May 1947 cover of *Mechanix Illustrated*. He rejected the idea because he saw no propellers or landing gear. Research shows that the navy built a single Flapjack and that it never flew outside the Bridgeport, Connecticut, area.

At the end of Fugate's report, he wrote, "Mr. Rhodes stated that he developed the negatives himself. He still had the negative of the first photograph (Exhibit III), but he could not find the negative for the second photograph."

On February 19, 1948, Lewis C. Gust, chief Technical Project Officer, Intelligence Department (the Project Blue Book files fail to identify the man or his organization beyond that), wrote what might be considered a preliminary report on the analysis of the photographs: "It is concluded that the image is of true photographic nature, and is not due to imperfections in the emulsion, or lack of development in the section in question. The image exhibits a 'tail' indicating the proper type of distortion due to the type of shutter used, the speed of the object and the fixed speed of the shutter. This trailing off conforms to the general information given in the report."

On May 11, 1948, Rhodes was again interviewed but this time by high-ranking people. Lieutenant Colonel James C. Beam, who worked with Colonel Howard McCoy, the head of intelligence at Wright field, and Alfred C. Loedding, who was a civilian employee at AMC and part of Project Sign, traveled to Phoenix. In their official report of their trip, they wrote, "Although Mr. Rhodes is currently employed as a piano player in a night club, his primary interest is in a small but quite complete laboratory behind his home. According to his business card, this laboratory is called

'Panoramic Research Laboratory' and Mr. Rhodes is referred to as the 'Chief of Staff.' Mr. Rhodes appeared to be completely sincere and apparently is quite interested in scientific experiments."

During the interview with Beam and Loedding, Rhodes mentioned that he did not believe that what he had seen was wind-blown debris. This is an obvious reference to Dr. Irving Langmuir's conclusion published in the "Project Grudge Report," that the object in the photographs could be "merely paper swept up by the winds."

In fact, that same Grudge report noted, "In subsequent correspondence to the reporter of this incident, the observer refers to himself as Chief of Staff of Panoramic Research Laboratory, the letterhead of which lists photography among its specialities. Yet, the negative was carelessly cut and faultily developed. It is covered with streaks and over a period of six months, has faded very noticeably."

The AMC opinion in the Project Grudge report, which followed Langmuir's statement about the possibility of windblown debris, was, "In view of the apparent character of the witness, the conclusion of Dr. Langmuir [that the photographs be discounted as paper swept up by the wind] seems entirely probably [sic]."

On June 5, 1952, now nearly five years after the pictures were taken, and before the massive publicity about UFOs was about to burst into the public consciousness, Colonel Arno H. Luehman, Deputy Director of Public Information wrote a letter to the Director of Intelligence concerning "Declassifying Photographs of Unidentified Flying Objects." In the first paragraph of the letter, he wrote, "This office understands that two photographs were taken by Mr. William A. Rhodes of Phoenix, Arizona, and that these photographs were turned over to Fourth Air Force Intelligence in July of 1947. This office has been contacted by Mr. Rhodes who is requesting return of his original negatives."

The letter continued, "The two photographs were copied by the Photographic Records and Services Division of the Air Adjutant

General's Office at this headquarters and are in a confidential file of Unidentified Missiles as A-34921AC and 34921AC."

On July 14, 1952, in still another letter, written by Gilbert R. Levy, we learn that the pictures and negatives were turned over to air force intelligence representatives at Hamilton Field on August 30, 1947. In that document, they are attempting to trace the course of the pictures from Rhodes to the FBI to army intelligence. What this suggests is that the Air Force wasn't sure of where the pictures and negatives were. They were attempting to shift the blame to others for the apparent loss of those pictures, including Rhodes himself.

In that same July 14 document, Levy noted that "a background investigation was run on Rhodes, by OSI, for the benefit of AMC, which reflected Rhodes had created the name PANORAMIC RE-SEARCH LABORATORY, to impress people with his importance. He was reported to be a musician by trade, but had no steady job. Neighbors considered him to be an excellent neighbor, who caused no trouble, but judged him to be emotionally high strung, egotistical, and a genius in fundamentals of radio. He conducts no business through his 'Laboratory,' but reportedly devotes all his time to research."

What all this information means is that Rhodes had surrendered his photographs and negatives to the government. And, although there is a suggestion that he knew where they were, that simply isn't borne out in the documents. Even the air force officers didn't know where the photographs were. That was why there were letters written from one office to another.

But, more important, there has been no real discussion about why the air force investigators labeled the case as a probable hoax. The discussion seemed to center around Rhodes's lifestyle. He didn't have a "real" job and had letterhead that labeled him as the chief of staff of his laboratory. None of that is a good reason for labeling the case a hoax. If that was all of their evidence against him, then it is fairly weak.

There is, however, one page of analysis of the photographs offered by John A. Clinton in the Project Grudge files. There is no clue, in the files, about who Clinton is. The analysis is not on a letterhead and there is nothing in the signature block to tell us anything about Clinton, his expertise, or why he was consulted about this particular case.

In the undated analysis of the photographs, Clinton wrote, "Preliminary analysis of the negative and prints leads me to doubt the story told by Mr. William A. Rhodes. Judging from the dimensions, the negative was exposed in a simple camera of the box type, which usually has a fixed focus (about ten feet), fixed shutter speed (about $\frac{1}{25}$ of a second) and a simple lens of the Meniscus type. Because of the above mentioned facts, it is unreasonable to assume that sharp outlines such as appears on the negative, could be secured from an object at 2,000 feet, traveling 400 to 600 mph. Furthermore, according to the story, the object (flying craft) was painted gray to blend in with the clouds. But, even if the object would be painted jet black, under the circumstances described, to obtain a contrast such as appears on the negative is also very doubtful. On all the prints, excepting the print marked "exhibit A", judging from the outlines, the object has a rotating motion (revolves around its center) instead of a forward motion, contradicting the version stated by Mr. Rhodes."

And that's all of the negative analysis of the photograph. Clinton, whoever he is, claimed the story told by Rhodes to be in conflict with that shown on the photographs. He assumed that the object was rotating, based on something he saw on the prints. Besides, Rhodes talked of the craft circling east of his house, moving north to south when he first saw it. That detail seemed to be an explanation for the conclusion drawn by Clinton that the object was rotating.

More important are the suggestions about the limitations of the camera used and the sharpness of the photographs obtained. Of course, if Rhodes, for whatever reason, overestimated the distance and the speed, then those problems might be resolved.

By 1952, when Rhodes was trying to get the pictures back, air force investigators, including the then-captain Dewey Fournet, suggested in a telephone conversation with the then-lieutenant Ed Ruppelt that "there is no information available as to whether or not Rhodes ever sent his negatives to the Air Force or whether he just sent prints. We do have some rather poor quality prints of the object. As you know, we have concluded that these photos were probably not authentic. It seems as if Mr. Rhodes attempted to get on the 'picture selling bandwagon' and if he can prove he sent the negatives to ATIC or to the Air Force and they were never returned, it may lead to a touchy situation."

So, five years after Rhodes's photographs were submitted, they were going to be rejected because the air force suspected he wanted to sell them. Without a single indication that such was the case, the air force rejected the photographs. There is no evidence anywhere that Rhodes ever sold the photographs nor is there any information in the Project Blue Book files to confirm he made a dime from the pictures.

There is one disturbing thing about the case, but it is not evident in the Blue Book file. In the mid-1960s, Dr. James E. McDonald corresponded with Rhodes about his case. McDonald wrote to Richard Hall, of NICAP (and later of the Fund for UFO Research), on February 18, 1967, that "I did a lot of checking on Rhodes degrees, because there seemed something odd about an honorary PhD based on the kind of work I could imagine him doing. Columbia said no record of any such degree. Geo. Washington said no record of a BA ever given to Rhodes in the period I specified. So I made a trip up there in December and spent an hour or so with him. Devoted most of my querying to the matter of the degree and his associations with inventory [sic], Lee De Forrest. . . . He [Rhodes] showed me a photo-miniature in plastic of the alleged Columbia degree, and he said he had the original somewhere in his files but did not show it to me. . . . As I kept going over the thing he finally volunteered the remark that he, himself, had checked with Columbia about a year after De Forrest presented him with the certificate, found no record of it, confronted De Forrest with the information, and was non-plussed by D F putting his arm over his shoulder and saying something to the effect, that, 'Well, my boy that's the way those things happen sometimes,' and saying no more about it. . . . But the fact that he lists himself in the Phoenix phonebook as Dr. Wm Rhodes in the face of that history constitues [sic] a cloud that would be impossible to overlook. Everything else checks out solidly in his story."

There isn't much else to be said about the case. The air force eventually changed its categorization of the Rhodes file from "Possible Hoax" to "Other (Hoax)." They had no other explanation for it and too often, when air force could find no plausible explanation, especially in cases of physical evidence such as this, they labeled it as a hoax. As we've seen, there simply is no justification for the label. Rhodes just fit no easy profiles, so it was easy to label him an eccentric and his case as a hoax.

But the real fact that remains is that no one ever showed that Rhodes's tale of seeing the craft was not as he described it. Analysis of the photographs left a great deal to be desired, but there was nothing in the photos that suggested a hoax.

The Chiles-Whitted Encounter:

JULY 24, 1948

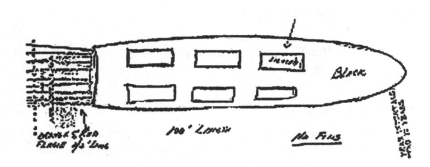

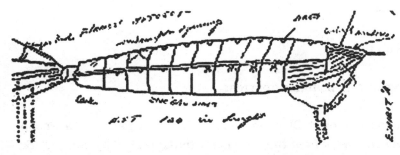

**The official drawings made by Chiles and Whitted just hours
after their sighting**

Location: Near Montgomery, Alabama

Witnesses: Captain Clarence S. Chiles, pilot
John B. Whitted, copilot

Craft Type: Cigar-shaped

Dimensions: One hundred feet long and as big around as the combined fuselages of three B-29s.

Primary Color: Red

Craft Description: Cigar-shaped with a double row of square windows

Sound: None

Exhaust: Long frame from rear

Sources: Project Blue Book files; *The Emergence of a Phenomenon: UFOs from the Beginning through 1957* by Jerry Clark;

Project Blue Book Exposed by Kevin D. Randle; *Watch the Skies* by Curtis Peebles; and *UFOs Explained* by Philip Klass

Related Sightings: Eureka, Utah, April 19, 1962; Midwest, March 3, 1968

Reliability: 9

Narrative: Clarence S. Chiles and John B. Whitted were flying a commerical DC-3 at about five thousand feet, on a bright, cloudless, and star-filled night. Twenty miles southwest of Montgomery, Alabama, they spotted what they thought was a jet slightly above them and to the right. Within seconds it was close enough that they could see a torpedo-shaped object, which had a double row of square windows.

Chiles directed the attention of his copilot to the object, saying, "Look, here comes a new Army jet job." The object approached in a slight dive, deflected a little to the left, and passed the DC-3 on the right, almost level to the flight path. After passing, it pulled up sharply and disappeared into a cloud.

Questioned within hours of the event by investigators, both men said that they believed the object was about a hundred feet long. Whitted said, "The fuselage appeared to be about three times the circumference of a B-29 fuselage. The windows were very large and seemed square. They were white with light which seemed to be caused by some type of combustion. I estimate we watched the object for at least five seconds and not more than ten seconds. We heard no noise nor did we feel any turbulence from the object. It seemed to be at about 5,500 feet."

Chiles, in a statement dated August 3, 1948, wrote, "It was clear there were no wings present, that it was powered by some jet or other type of power shooting flame from the rear some fifty feet. . . . Underneath the ship there was a blue glow of light."

Apparently all the passengers in the DC-3 were asleep with the exception of Clarence L. McKelvie. Chiles wrote, "After talking to the only passenger awake at the time, he saw only the trail of fire as it passed and pulled into the clouds."

Within hours of the sighting, Chiles and Whitted were interviewed on radio station WCON in Atlanta, Georgia. They were also interviewed by William Key, a newspaper reporter. At some point during one of the interviews, someone suggested they had been startled by a meteor, but both men rejected the idea. Both pilots had seen many meteors during their night flights and were aware of what they looked like and how they performed.

There are some other points to be made. Chiles and Whitted were reported as believing that a disturbance of their plane was caused by the UFO. In a newspaper article written by Albert Riley, Riley quotes Chiles as saying, "Its [the craft's] prop-wash or jet-wash rocked our DC-3." In another article that is part of the Blue Book files, Chiles and Whitted's alleged assessment is again cited: "Both reported they could feel the UFO's backwash rock their DC-3."

According to the interviews with newspaper reporters, then, Chiles and Whitted were supposed to have said that they had felt turbulence that they believed was the result of the passage of the object. If true, that single fact would rule out a meteor as the UFO.

However, a search of the Blue Book files reveals that in a statement he signed on August 3, 1948, Chiles said, "There was no prop wash or rough air felt as it [the object] passed."

In a statement taken by military officers in the days that followed the sighting, Whitted said, "We heard no noise nor did we feel any turbulence from the object."

It would seem, then, that neither man reported any turbulence or disturbance of the air as the object passed them. The quotes from the newspapers, therefore, seem to be in error.

Dr. J. Allen Hynek, the scientific consultant to the air force's Project Blue Book, was asked for his assessment of the case. He could find no "astronomical explanation" if the case was accepted at face value. In other words, Hynek was saying that if the testimony of Chiles and Whitted was accepted, then it couldn't be explained. He also wrote, "[The] sheer improbability of the facts as stated . . . makes it necessary to see whether any other explanation, even though far-fetched, can be considered."

What this commentary does is provide us with a couple of conclusions to this case. First, Captain Robert R. Sneider wrote on November 12, 1948, "A preponderance of evidence is available to establish that in almost all cases an unidentified object was seen within stated times and dates over an extended area, pursuing a general Southerly course. Descriptions as to size, shape, color and movements are fairly consistent."

Sneider also wrote, "The flying anomaly observed, remains unidentified as to origin, construction and power source."

On July 13, 1961, Dr. Donald E. Menzel, wrote to Major William T. Coleman at the Pentagon, discussing UFO sightings and a book that Menzel was writing. Menzel noted, "One further question that we have. Our study of the famous Chiles case indicates that the UFO was merely a meteor. Apparently this was a considered solution in the early days. We wonder why it was abandoned."

Of course, Menzel is referring to Hynek's suggestion that had not gained much support at Sign in 1948. By way of contrast to Menzel's argument for the meteor theory is Dr. James McDonald's counterargument, which was based on his review of the Blue Book files and his own personal interviews with Chiles and Whitted. McDonald wrote, "Both pilots reiterated to me, quite recently, that each saw square ports or windows along the side of the fuselage-shaped object from the rear of which a cherry-red wake emerged, extending back 50–100 feet aft of the object. To term this a 'meteor' is not even qualitatively reasonable. One can reject testimony; but reason forbids calling the object a meteor."

So we're back to where we started. Two airline pilots see something flash through the sky at them. Both talk of a double row of square windows, a cigar shape, and a red flame from the rear. A passenger on the plane sees a streak of light, but no details.

And, an examination of the case reveals no persuasive evidence to suggest that a meteor is, in fact, responsible for the sighting. If, as Sneider, one of the air force investigators on the case suggested, we take the sighting at face value, then contrary to the air force opinion, there is no solution for the sighting. It should have stayed categorized as unknown or unidentified.

110

One of the Best Ever:

MAY 11, 1950

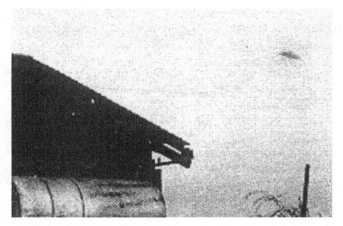

What might be the best of the UFO photographs taken in the early 1950s

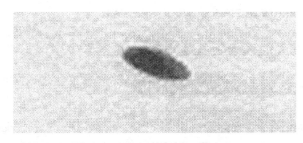

Close-up of the object seen by Paul Trent and his wife

Location: McMinnville, Oregon

Witnesses: Mr. and Mrs. Paul Trent

Craft Type: Daylight disc

Dimensions: Thirty feet in diameter

Primary Color: Gray

Sound: None

Exhaust: None

Quality of Photos: Sharp and clear

Number of Photos: 2

Type of Camera: Unavailable

Sources: Project Blue Book files; *Project Blue Book Exposed* by Kevin Randle; *The UFO Encyclopedia* by Jerry Clark; *UFOs*

Explained by Philip Klass; *Scientific Study of Unidentified Flying Objects* edited by Daniel S. Gillmar

Reliability: 10

Narrative: According to the witnesses, Mr. and Mrs. Paul Trent, who lived on a farm near McMinnville, Oregon, they were confronted by a large, slow-moving, disc-shaped object traveling toward them from the northeast. Mrs. Trent was outside, having just fed the rabbits and was the first to see the object. She called to her husband, who left the house, watched the object, and then ran back into the house to grab a camera. Mr. Trent then took two pictures of the object as it flew over. As he was taking the pictures, Mrs. Trent noticed that her in-laws were on the back porch of their home about four hundred feet away. She shouted, but they didn't hear her. She then ran into the house to call her in-laws on the phone. Because her mother-in-law went to answer the telephone, she didn't see the object, but Mrs. Trent's father-in-law did glimpse it as it disappeared into the west.

According to witness statements made many years after the event, Mr. Trent said that he took the first picture, wound the film, which in 1950 was a time-consuming process, and took the second. By then, the object, which had been moving slowly, began to accelerate.

The Trents described the object as very bright and silvery. Both saw a superstructure on top of the craft, which appeared in the photographs. William K. Hartmann reported that the Trents told him that it resembled a "good-sized parachute canopy without the strings, only silvery-bright mixed with bronze." Hartmann noted in his report that the "rather bright, aluminum-like, but not specular, reflecting surfaces appears to be confirmed by analysis of the photos. . . . There was no noise, visible exhaust, flames, or smoke."

Although they had what could be the first authentic pictures of an extraterrestrial craft, the Trents didn't develop the film immediately because there were still frames to be used. They did, however, according to Jerry Clark, discuss the sighting with friends and family. When the pictures were developed, and after Mr. Trent had mentioned the incident to his banker, Frank Wortmann, copies of the photos were put on display in the bank.

Reporter Bill Powell saw the pictures in the bank, interviewed the Trents, and convinced them to allow him to publish the photos. Mr. Trent said that he was afraid that they would get into trouble with the government, but Powell convinced him otherwise.

On June 8, about a month after the pictures were taken, Powell's story appeared in the *McMinnville Telephone Register.* He had examined the negatives, found on the floor where the Trent's children were playing with them, and decided there was no evidence of tampering or of a hoax.

Once the pictures were published, the Trents found themselves the focus of national attention. *Life* borrowed the negatives from Powell and printed them in the June 26 issue. Although the Trents were promised the return of their negatives, it would be seventeen years before they received them. Before the photos were given back, they were loaned to Hartmann for his study, which he conducted under the auspices of the Condon Committee.

Hartmann made his own investigation of the Trents' tale and examined the negatives. It was because of the Condon Committee that the negatives were located and eventually returned to the Trents.

Hartmann, knowing that the photographs were only as good as the reputation of the photographer, noted that a number of local residents had come forward to "attest to the witnesses' veracity. They appear to be sincere, though not highly educated or experienced observers."

More important, because of criticism that would be leveled by later investigators, Hartmann wrote, "During the writer's interview with them [the Trents], they were friendly and quite unconcerned about the sighting."

Hartmann, in his evaluation of the case, also wrote, "Two inferences appear to be justified: 1) It is difficult to see any prior motivation for a fabrication of such a story, although after the fact, the witnesses did profit to the extent of a trip to New York; 2) it is unexpected that in this distinctly rural atmosphere, in 1950, one would encounter a fabrication involving sophisticated trick photography (e.g. a carefully retouched print). The witnesses also appear unaffected by the incident, receiving only occasional inquiries."

It can also be noted that a few months later, when Nick Mariana made his brief film of two objects he saw over Great Falls, Montana, there was no economic benefit. He lost sponsorship on many of the radio stations that had used his sportscasts. Even in 1950 there was a feeling that those who saw and reported UFOs were unsophisticated, uneducated, and probably a little off center in their thinking.

Hartmann continued his analysis of the Trent photographs by writing, "As stated previously, it is unlikely that a sophisticated 'optical fabrication' was performed. The negatives have not been tampered with."

He concluded his section of the report by writing, "This is not of the few UFO reports in which all factors investigated, geometric, psychological, and physical appear to be consistent with the assertion that an extraordinary flying object, silvery, metallic, disk-shaped, tens of meters in diameter, and evidently artifical, flew within sight of two witnesses. It cannot be said that the evidence positively rules out a fabrication, although there are some physical factors such as the accuracy of certain photometric measures of the original negatives which argue against fabrication."

Let's take just a moment and think about this. Many of the scientists who have commented on the UFO phenomenon have asked, repeatedly, Where is the physical evidence? Here is an example of that evidence, examined by a scientist who has suggested, at the very least, that the Trent pictures show a large manufactured craft and there is no indication of a hoax. Shouldn't this evidence, even though there are but two primary witnesses, demand that an impartial and scientific investigation of UFOs be undertaken? Isn't this an example of that physical evidence demanded by science?

It is also interesting to note that the Trents, as well as the reporter, Powell, claimed to have been visited by plainclothes representatives of the air force and members of the FBI. Powell told Dr. Bruce Maccabee, a Navy physicist and member of FUFOR (Fund For UFO Research), that two weeks to a month after his newspaper story, a plainclothes agent of the air force showed up at the *Register* office and demanded the pictures. Repeated appeals by Powell to recover the prints went unanswered.

The claims made by Powell and the Trents are interesting because the air force file on the McMinnville case is labeled as "Information Only." In a letter dated March 10, 1965, Lieutenant Colonel John P. Spaulding responded to an inquiry from W. C. Case. He wrote, "The Air Force has no information on photographs of an unidentified flying object taken by Mr. and Mrs. Trent of McMinnville, Oregon. In this regard, it should be noted that all photographs submitted in conjunction with UFO reports have been a misinterpretation of natural or conventional objects. The object in these photographs have a positive identification."

Is it necessary to point out the contradiction in the letter? The air force has no information about the sighting or the photographs, but the photographs have a positive identification.

None of this information is, of course, the last word to be heard on the Trent photo case. Philip Klass, in *UFOs Explained,* found a number of reasons to believe the photographs were a hoax. All other possible mundane explanations had been rejected either by the witness testimony or by the photographs themselves. Clearly they did not show balloons, aircraft, meteors, ball lightning, insects, processing flaws, the moon, stars, atmospheric phenomena, astronomical phenomena, paper blowing in the wind, or any of the other misinterpretations that have been used to explain UFO sightings. There were only two possible explanations: Either the pictures were a hoax or they were of a large manufactured craft of unknown origin.

Klass launched his own investigation of the photographs, obtaining prints from the original negatives that he passed along to his colleague Robert Sheaffer. While Sheaffer studied the pictures, Klass began "to probe for 'soft spots' in the Trents' story of events. . . ."

Klass reprinted the statement that Mrs. Trent had made to the newspaper in 1950. She said, "*It was getting along toward evening*—about a quarter to eight. We'd been out in the back yard. *Both of us saw the object at the same time.* The camera! Paul thought it was in the car but I was sure it was in the house. I was right—the Kodak was loaded with film. Paul took the first picture. The object was coming toward us and seemed to be tipped up a bit. It was very bright—almost silvery—and there was no noise or smoke [emphasis added by Klass]."

Klass then makes an assumption with which we don't agree. He noted that the Trents were aware of the UFO phenomena and had read enough about it so that they would have recognized the importance of the pictures. But, rather than wasting the last three exposures on the film, they waited to finish the roll before having it processed. He was also surprised that after having the film developed, the Trents did nothing with them except show them to friends. Klass believed that if they had actually photographed what they believed to be a flying saucer, they would have been more excited about it, told more people about it, and would have realized the financial potential of having the first authentic pictures of an alien spacecraft.

It could be argued that the Trents, if familiar with the UFO literature, would also know how those reporting UFOs were treated. It made no difference to some that evidence was available. They still believed people who saw UFOs were somehow less than reliable. Within days of Kenneth Arnold's sighting in June 1947, newspapers were suggesting that those who reported flying saucers were drunks. One newspaper reported that the "flying disks" had been seen in thirty-eight states, except Kansas, which was dry. The answer to Klass's conjecture could be simply that the Trents didn't want to be seen as "drunks" or "nuts."

117

Even after seeing the pictures, the Trents still did nothing until persuaded to do so by friends. Mr. Trent told Bill Powell that he was afraid that he would get into trouble with the government. Klass points out that Trent would get into as much trouble in May as he would in June. He is suggesting that the reason for the delay makes no sense. Of course, it could be argued that Mr. Trent's friends talked him into releasing the pictures, something he would not have done on his own.

In his search for "soft spots" Klass does find one point that is somewhat significant. The Trents are repeaters. That means they had seen UFOs on more than one occasion. Klass noted that NICAP, at one time the most respected of the civilian UFO organizations, warned that one reason to reject photographic evidence was if it came from witnesses who had seen UFOs more than once. This is, of course, a subjective opinion, and not relevant to the discussion of the Trent photographs.

More important, and of much more relevance, are the findings of Robert Sheaffer. According to Klass, Sheaffer had detected what he believed to be the fatal flaw in the Trent pictures. Klass wrote, "Sheaffer's keen eye noted that there were distinct shadows on the *east* wall of the Trent garage, caused by the overhanging eaves of the roof. This indicated that the pictures had been *taken in the morning,* and not shortly after sunset as the Trents had claimed. Sheaffer used his training in astronomy and mathematics to calculate the time of day when the photos were made, based on the position of the shadows. He concluded that the pictures had been taken at approximately 7:30 A.M., not 7:45 P.M. Furthermore, his analysis indicated that the photo which the Trents claimed had been taken first had really been shot *several minutes after the other picture, and not a few seconds earlier as the Trents said* [emphasis in the original]."

If this assessment was true, then the Trents had been caught in a lie. If they had lied about one aspect of the case, then the whole thing would collapse. But why lie about the time of day? It would make no difference to researchers if the pictures were taken in the morning or the evening.

Hartmann, according to what Klass wrote, had also noted the shadows but could think of no reason for the Trents to lie about the time. Klass, however, found what he believed to be the reason. He wrote, "Most farmers are out working in the fields around 7:30 A.M., when the photos were actually taken, and Trent's neighbors would think it odd that they too had not seen the giant UFO that the Trents claimed to have seen and photographed. But to claim that the pictures were taken at 7:45 P.M., when most farmers have retired to their houses for dinner, would eliminate most potential witnesses who might dispute the story and photos."

But that really makes no sense. Farmers are in their fields at all hours, and it isn't surprising to see them working long after dusk, which is why there are lights on their equipment. More important, more of the general population is outside in the evenings. Even the most cursory examination of sighting reports shows that more are reported between six in the evening and midnight. If the Trents were motivated to move the timing of the sighting, it would be to move it into the morning when there were fewer opportunities for other witnesses.

And, if the timing, as provided by the Trents is accurate, then Sheaffer's suggestion about the ordering of the photographs is reversed. In other words, if the pictures were taken in the evening, then they were taken in the order that the Trents claimed.

It is appropriate to point out that both Klass and Sheaffer have written books explaining away all UFO sightings. Neither Klass nor Sheaffer believe that there have ever been any legitimate UFO sightings and that belief has colored their thinking. If there are no UFOs, then there must be some fact, some bit of information suggesting that the case is explainable or is a hoax.

In fact, if we examine the statements made by the Trents, we do see minor changes in the timing and the events. These are not the result of a hoax but of the problems of human memory. Klass is always quick to point out that memories fail and information is related inaccurately. Here, however, the Trents are expected to remember the exact nature of all the events, including minor and often inconsequential details. These changes are to be expected and are not indictative of a hoax.

The shadows under the eaves, which Sheaffer believed indicated the photographs were taken in the morning, have been written off as random light scattering. Dr. Bruce Maccabee, a physicist for the navy, and who has made a number of studies of photographs, found nothing inconsistent with the story told by the Trents and the information available in the photographs.

Hartmann, in his study, had also noticed the shadows under the eaves, and ignored them. He found nothing to suggest that the Trents were not telling him the truth, and no reason for them to change the time of the sighting. His investigation suggested that the story and the photographs were a consistent whole package.

The point here, however, is not to determine if the photographs are hoaxes or of a real, large, manufactured object. The point is to determine if there is any physical evidence that UFOs might be real. The Trent photographs provide us with that sort of evidence. We have two main witnesses, the possibility of a third, and we have the photographs, which have been authenticated, but characterized to be a hoax by men with the credentials and the expertise to make those sorts of claims.

It can be noted that the personal bias of some of those men is showing. As noted earlier, both Klass and Sheaffer believe there are no UFOs, which could color their thinking. But Maccabee is a believer. That could also color his thinking. That leaves us with Hartmann, who seemed to be a disinterested third party, even though he worked with the thoroughly skeptical Condon Committee.

But Hartmann had made statements on both sides. In the Condon Committee report, he suggested that the photographs did show a large manufactured object and he found no evidence of a hoax. Klass reported that when he confronted Hartmann with the facts of his investigation, Hartmann changed his mind. Klass wrote, "It is a tribute to Hartmann that when Sheaffer and I presented him with the results of our investigations, he promptly revised his earlier views on the Trent pictures. Hartmann seemed especially impressed with Shaeffer's efforts in demonstrating that the pictures had been taken around 7:30 A.M. 'I think Sheaffer's work removes the McMinnville case from consideration as evidence for the existence of disklike artificial craft,' Hartmann said, using his favorite euphemism for extraterrestrial spacecraft."

But neither Klass nor Sheaffer could offer a good reason for the changing of the time of the sighting. And if the timing was accurate, and the shadows under the eaves were due to random-light scattering, then we are back where we began, which is that we have two interesting photographs that suggest the UFO phenomenon deserves a closer look.

The Lubbock Lights:

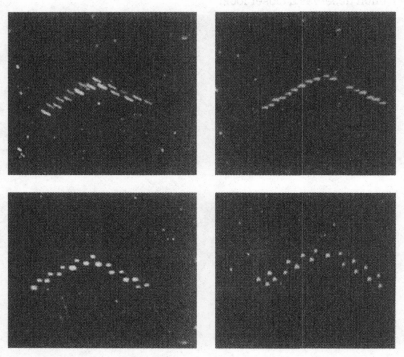

The four photographs taken by Carl Hart, Jr., have not been satisfactorily explained.

Location: Lubbock, Texas

Witnesses: Carl Hart, Jr.

Craft Type: Nocturnal light

Dimensions: Thirty to sixty feet in diameter

Primary Color: White

Craft Description: White lights

Sound: None

Exhaust: None

Quality of Photos: Sharp and clear

Number of Photos: 5 (though only four have survived)

Type of Camera: Unavailable

Sources: Project Blue Book files; *Lubbock Lights* by David Wheeler

Relation to Other Sightings: Salem, Massachusetts

Reliability: 8

Narrative: The Lubbock Lights story began on a hot August night as several professors from Texas Tech College (later University) sat outside. A group of dully glowing lights flashed overhead. They moved silently, crossed the sky rapidly, and seemed to be in some kind of loose, but organized, formation. They were only in sight for two or three seconds, and none of the professors got a very good look at them.

The professors W. I. Robinson, A. G. Oberg, and W. L. Ducker discussed what they had just seen, trying to figure out what it might have been. They also tried to determine what to do if the lights returned. An hour or so later, the lights reappeared, and this time the professors were ready.

The lights were softly glowing bluish objects in another loose formation. It seemed to the professors that the first group had been in a more rigid and structured formation than the later groups.

To the professors, the next logical move was to learn if anyone else had seen the objects. Ducker called the local newspaper and spoke to the managing editor, Jay Harris, who wasn't interested in the report. Ducker, however, convinced him that a story should be printed. Harris finally agreed but only if Ducker allowed his name to be used. Ducker refused.

However, a few minutes later, Ducker called back and agreed. In fact, he said that Harris could print the names of all the professors, but only if Harris called the college public-relations department and cleared it with them.

The newspaper story was successful in finding other witnesses: It was reported that there were others who claimed to see the lights that same night. That seemed to be some corroboration of the lights seen by the professors. But the important sighting, at least in the minds of the air force officers who later investigated the case, was made by Joe Bryant of Brownsfield, Texas.

Bryant told air force officers that he was sitting in his backyard when a group of the dim lights flew overhead. He described them as having a "kind of a glow, a little bigger than a star." Not long after that, a second group appeared. Neither of the groups was in any sort of a regular formation, a detail that the air force chose to ignore.

There was a third flight, but instead of flying over Bryant's house, they dropped down and circled the building. As he watched, one of them chirped, and Bryant recognized them immediately. He identified them as plovers, a bird common in west Texas. When he read the account of the professors in the newspaper the next day, he knew immediately what they had seen. If he hadn't been able to identify the last flight, if one of the birds hadn't chirped, he would have been fooled too.

The professors, unaware of what Bryant had seen and believed, set out to obtain additional information. Joined by other professors and professionals, including Grayson Meade, E. R. Hienaman, and J. P. Brand, they equipped teams with two-way radios, measured a base from the location of the original sightings, then they staked out the area. They hoped for additional sightings along the base line. Knowing the length of that line, the time of the sighting, and the location and direction of flight, they would be able to calculate a great deal of important and useful information that might tell them what they had been seeing.

The problem was that none of the teams ever made a sighting. On one or two occasions, the men's wives, who had remained at one house or the other, said that they had seen the lights, but the men at the bases saw nothing. The plan of calculating the data fell apart.

Then, on August 31, the case took an amazing turn. Carl Hart, Jr., a nineteen-year-old amateur photographer, managed to take five pictures as the lights flew over his house in the middle of

Lubbock. Lying in bed at about ten o'clock, he saw the lights flash over. Knowing that they sometimes returned, he prepared for that. When the lights appeared a few minutes later, he was ready, snapping two pictures of them. Not long after that, a third group flew, and he managed to get three additional pictures.

Harris, the Lubbock newspaper editor, learned about the pictures when a photographer who worked for him periodically called to tell him that Hart had used his studio to develop the film. Harris, the ever-reluctant newsman, suggested that Hart should bring the pictures by the office.

Naturally the newspaper feared a hoax. Harris and the newspaper's lead photographer, William Hams, talked to Hart on a number of occasions over the next several hours. Harris bluntly asked if the pictures were faked. Hart denied it. When Kevin Randle spoke to Hart about the incident forty years later, he asked Hart what he had photographed. Randle didn't want to accuse him of faking the pictures, but he wanted to know if Hart had changed his mind with the passage of years. Hart told him that he still didn't know what he had photographed.

Hams later decided to try to duplicate Hart's pictures. From the roof of the newspaper office, he attempted to photograph, at night, anything that flew over. He thought that if he could duplicate the pictures, he would be able to figure out what they showed. He waited, but all he saw was a flight of birds that were barely visible in the glow of the sodium vapor lamps on the streets below him. The birds were dimly outlined against the deeper black of the night sky and flew in a ragged V formation.

Hams took photographs of the birds, but when he developed the film, the images were so weak that he couldn't make prints. He repeated his experiment on another occasion but was no more successful. From his experience, he was convinced that what Hart photographed couldn't have been birds under any circumstances.

Air force investigations were conducted throughout the fall of 1951. Investigators were dispatched from Reese Air Force Base on the west side of Lubbock. They spoke to Hart on a number of occasions. They forwarded copies of their reports to both Project Blue Book headquarters and to the Air Force Office of Special Investigation headquarters in Washington, D.C. Ed Ruppelt even made a trip to Lubbock to speak to the witnesses, including Carl Hart.

During those interviews, Hart was advised of his rights under the Constitution of the United States. The investigators were playing hardball with the teenager. They were trying to pick apart his story to prove that he had somehow faked the pictures. Between November 6 and 9, during still another investigation of the Lubbock Lights, Ruppelt and AFOSI Special Agent Howard N. Bossert again interviewed Hart. In their report, they wrote, "Hart's story could not be 'picked apart' because it was entirely logical. He [Hart] was questioned on why he did certain things and his answers were all logical, concise, and without hesitation."

When Kevin Randle talked to the experts at Texas Tech about the possibility of birds, Loren Smith told him that there are ducks in the Lubbock area that fly in V formations but that they are reddish-maroon and have no white on them to reflect the lights. Although migratory birds do fly past Lubbock, it is later in the year. What this information means is that there were no birds in the area that account for the photographs.

What we must do is separate the Hart photographs from the rest of the Lubbock case. In fact, we must look at all the sightings individually, realizing that a solution to one is not necessarily the solution to another or to all the reports.

First we have the sightings made by the professors. Clearly this was something that was unusual. They were unable to identify the lights. They then, using their scientific training, set about to find out what they had seen. Although their plan was good, the phenomenon did not cooperate with them. There were some facts obtained, and these can lead us to some conclusions.

For example, the professors had originally estimated the objects as being very large and flying at a very high altitude. When

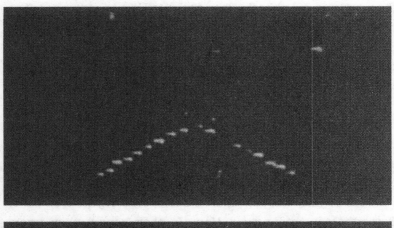

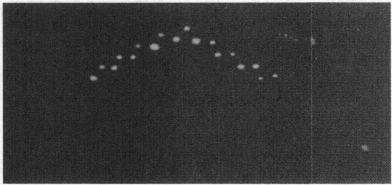

The plover flies in groups of seven or eight and does not fly in V formations, according to local experts.

they established their base lines, they never saw the objects again. The wives, however, reported the objects overhead. That would seem to indicate that the objects were smaller and much lower than originally thought. In fact, it suggests that they were *much* smaller and *much* lower. The door is open for the birds, though the problem, once again, is the lack of an appropriate bird appearing in the Lubbock area.

Or is the door still open? Joe Bryant claimed that he saw the lights too, but that one of them, or several of them, swooped out of the sky to fly around his house. At that point he identified them as plovers.

From Bryant's claim, the air force investigators extrapolated that all the Lubbock sightings could be explained by birds. In one of the reports, the investigators wrote, "It was concluded that birds, with street lights reflecting from them, were the probable cause of these sightings. . . . In all instances the witnesses were located in an area where their eyes were dark-adapted, thus making the objects appear brighter."

The problem is, and one with which the air force investigators never dealt, was that similar sightings, that is, strings of lights in the night skies, were seen all over west Texas. From as far north as Amarillo to as far south as the Midland-Odessa area, reports of these sorts of sightings were made. Birds and the newly installed sodium-vapor lamps in specific areas of Lubbock do not provide an adequate explanation.

What is relevant here, however, is that air force officers made a long, complex investigation of the sightings. Ruppelt flew down from Wright-Patterson Air Force Base, and officers and investigators were dispatched repeatedly from nearby Reese. They actually spoke to the witnesses in person, searched for evidence, analyzed the photographs, and conducted follow-up interviews. Ruppelt made it clear that he believed there was a plausible, mundane explanation for the sightings, but never officially said what it was. Later, by searching his files, Kevin Randle learned that Ruppelt thought, personally, the Lubbock Lights were explained by fireflies.

Of course, that explanation didn't explain the photographs. Ruppelt wrote that he never found an explanation for them. "The photos were never proven to be a hoax but neither were they proven to be genuine." According to Ruppelt, "There is no definite answer."

Coast Guard Formation:

JULY 16, 1952

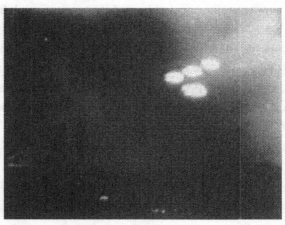

UFO formations photographed by Coast Guard personnel

Location: Salem, Massachusetts

Witnesses: Shell Albert, Thomas Flaherty

Craft Type: Luminous disc

Dimensions: Thirty feet in diameter

Primary Color: White

Craft Description: White lights

Sound: None

Exhaust: None

Quality of Photos: Blurred and indistinct

Number of Photos: 1

Type of Camera: Unavailable

Sources: Project Blue Book files

Relation to Other Sightings: Lubbock Lights

Reliability: 6

Narrative: A photographer, at work in a building at the Coast Guard installation in Salem, Massachusetts, saw four oval objects hovering over a parking lot outside. He grabbed a camera from a nearby table and moved toward the window. As he prepared to pull the side from the camera, he noticed that the objects had dimmed. At that time he assumed that he was seeing some kind of reflection.

He then ran out of the room, looking for another witness. He found a base-hospital man, who accompanied the photographer back to the room. As he entered, the photographer noticed that the lights were brilliant again. He moved to the window, and the lights vanished.

Investigation by the air force, based on the stories told by the photographer and the hospital man, suggested that the lights were reflections of internal lights on the dirty windows.

The air force investigation deemed the following points pertinent: The camera was set on infinity, meaning everything beyond a certain range would be in focus, and in the photograph, the cars in the parking lot are sharply focused, but the lights are blurry and out of focus. All four lights have the same general configuration and outline in spite of the blurring.

The air force also set up a number of bright spotlights so that they would be visible from the window where the photo was taken. Whenever the spotlights could be seen from the window, they produced reflections on the cars in the parking lot. Since the photograph of the objects showed no such reflections, the conclusion was that the lights were between the photographer and the lot, meaning that they almost had to be reflections on the dirty glass windows.

The Stock Domed Disc:

Location: Passaic, New Jersey

Witnesses: George Stock and his father

Craft Type: Domed disc

Dimensions: Thirty feet in diameter

Primary Color: Gray

Craft Description: Silver-gray, domed disc

Sound: None

Exhaust: None

Quality of Photos: Sharp and clear

Number of Photos: 8

Type of Camera: Unavailable

Sources: Project Blue Book files

Reliability: 3

Narrative: The official air force investigation into the stock disc sighting began in late November, 1952, when the officers assigned to Project Blue Book finally had the opportunity to look into the case. According to Charles Gregg, a staff writer for the *Herald-News* in Passaic, a man came into the office on the afternoon of July 31, 1952, with a series of pictures of a domed disc he had taken earlier in the day. The man, identified by some only as Riley, claimed he was visiting a friend and was talking to him in front of the house when the object appeared, traveling "southwest at a leisurely speed. As it drew nearer it came to a stop overhead for a few minutes, about two hundred feet above the ground."

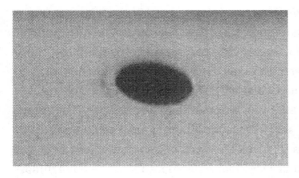

Stock took one of the first series of pictures that showed the classic domed disc.

He described the disc as about thirty feet in diameter and grayish in color. It made no sound. Before taking off again, it tilted, "as to observe the ground. An antenna, or something like that darted out of the dome's top for a moment and was withdrawn." The saucer then disappeared.

Gregg, in a letter to the air force, said that Riley claimed there were other witnesses, but he hadn't the time to look for them. Riley wouldn't leave the negatives, and Gregg believed that he was taking them to other newspapers to sell.

Later that night of July 31, Riley appeared at another newspaper, and a man named Dixon called the Air Technical Intelligence Center at Wright-Patterson AFB, Ohio, to let the air force know about the sighting. Major Herman called Dixon back and asked about "the photograph of an aerial phenomena."

Dixon responded, "That's about your way of putting it." He then went on to say, "Briefly, the story is this. A commercial photographer came in with the story that he was visiting a friend when the two of them saw the saucer overhead. The guy grabbed the friend's fixed focus camera and grabbed a half a dozen shots before the saucer moved away. He wanted to sell us the pictures. The pictures are phenomenally clear for anything of the sort. . . . What it looks like, off-hand, is something very close to the camera because it is fairly sharp considering the type of camera used. On the other hand, the man swears, and he has one witness who will back him up, that it is a legitimate photograph . . . we have threatened him with jail and all that and he still insists that it was correct."

Dixon and Herman discussed the photographs for a few minutes more. Dixon reported that no one really knew the man so that he couldn't say whether or not he was credible. Dixon did say that he thought the pictures were too good to be true and was inclined to believe that they were a hoax.

After several minutes, Captain Edward Ruppelt entered the conversation. All three men, although neither Herman nor Ruppelt had seen the pictures, were afraid they were being duped.

Ruppelt decided that it would be a waste of time to pursue the matter any further, based on what Dixon had said.

Even though they felt the sighting was a poor case, they did initiate an investigation into the background of Riley. They learned that he ran a small commercial film-processing business and that he didn't take pictures for a living. The people interviewed had nothing derogatory to say about Riley. The case was closed.

And then reopened. It was learned that the man who had sold the pictures was not, in fact, the man who had taken them. On November 19, 1952, an air force investigator interviewed George Stock, after it was confirmed that he was the photographer.

The story told by Stock was similar to the one told by Riley. Stock had been outside about 10:15 on the morning of July 31, when he saw the object overhead. He ran into the house for a camera, shouting to his father, "I think I see a flying saucer."

Both men rushed into the backyard. Stock noticed the saucer was closer, and he began taking pictures of it. He managed to get seven before the object disappeared to the southeast.

Air force investigators learned that the sighting lasted between five and seven minutes. The sky was clear, and the temperature was in the mid-eighties. There were only two witnesses, Stock and his father. Riley, the man who had tried to capitalize on the photographs, hadn't been there and only had the facts as Stock related them.

Late on the afternoon of July 31, Stock took the pictures to Riley for processing. Stock stood by while Riley developed the film and then took the pictures and negatives, but not before Riley had made an extra set of prints, which he had tried to sell.

Subsequent investigation by the Air Force Office of Special Investigation showed that both Stock and his father were considered to be reliable men. The neighbors said that neither had ever participated in any type of a hoax.

Air force investigation led to the Federal Telephone and Radio Corporation in Clifton, New Jersey, which makes radar and microwave equipment. A check with the corporation president

THE STOCK DOMED DISC

showed that no experiments were being conducted on the day of the sighting.

In the end, the air force could find nothing in Stock's background to suggest that he might have engineered the sighting. The AFOSI agent's report claimed that both Stock and his father were "honest, trustworthy and loyal American citizens."

The sighting, while questioned at first because of the clarity of the pictures, was unsolved. The Stocks were honest, and the negatives, five of which were examined by the air force, showed no signs of tampering. However, without corroborating evidence, investigators were left with a sighting that was very good but without the body of proof needed to make it a great sighting.

The Muskogee Disc

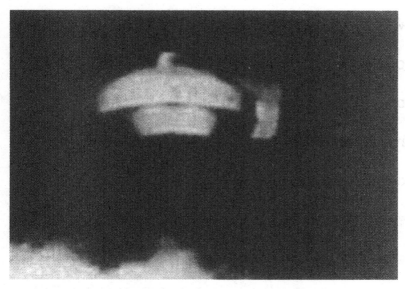

A craft that would become famous near Gulf Breeze, Florida, more than two decades later

Location: Muskogee, Oklahoma

Witnesses: Unknown

Craft Type: Disc

Dimensions: Seventy-five to eighty feet in diameter

Primary Color: Gray

Craft Description: Disc-shaped

Sound: None

Exhaust: None

Quality of Photos: Sharp and clear

Number of Photos: 1

Type of Camera: Super Ikonia

Sources: Project Blue Book files; APRD files

Relation to Other Sightings: Gulf Breeze, Florida 1987

Reliability: 2

Narrative: The official air force investigation began in 1954 and originated from the Eleventh OSI District at Tinker Air Force Base outside of Oklahoma City. According to the files, a postal employee said that after lunch sometime during the summer of 1953, he had still been near the restaurant where he'd had his meal when he spotted an object glinting in the sun. He had his camera with him and took a single photograph.

The witness said that he believed the object was traveling at about three to five hundred miles an hour and that it was about six

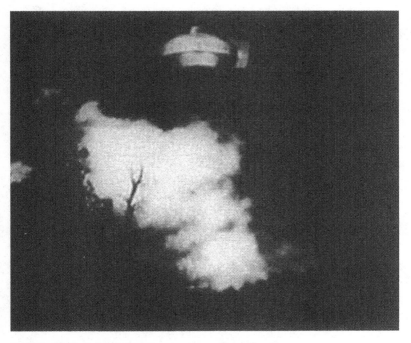

A high-flying UFO that was gone in a flash

or seven thousand feet above the ground. He thought it was seventy-five to eighty feet in diameter, was traveling in a southerly direction, and that it made no noise as it moved. He said that he had it in sight for about three seconds.

According to the witness, he didn't mention the sighting to anyone because he didn't want to get involved. Only reluctantly did he forward the information to the air force. He didn't want to be officially interviewed, according to the air force files.

The air force concluded that the sighting was a hoax. They found it odd that the object was centered in the picture and that there was no "lateral blurring." In their opinion, either the object itself, if moving as fast as claimed, or the trees in the foreground should have been blurred. This observation suggested to the air force investigators that the report disagreed with the physical evidence. Hence, to them, the "photo and incident are of doubtful validity."

Hoax or Reality?:

MAY 15, 1955

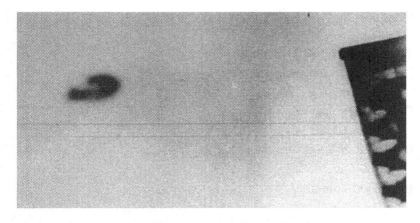

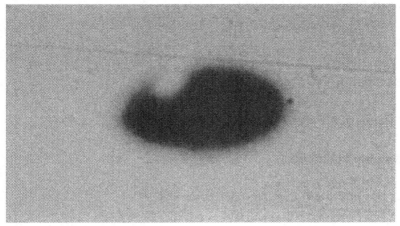

Location: New York, New York

Witnesses: Warren Siegmond
               Jeannine Bouiller

Craft Type: Disc

Dimensions: Precise dimensions not available.

Primary Color: Gray

Craft Description: Elliptical

Sound: None

Exhaust: None

Quality of Photos: Sharp and clear

Number of Photos: 5

Type of Camera: English reflex-type

144

Sources: Project Blue Book files; *Look* ("Flying Saucers," Special one-time issue, 1966)

Reliability: 3

Narrative: Warren Siegmond had rented a camera a couple of days before the sighting, and on the afternoon of May 15, 1955, at about four o'clock, suggested that he take some pictures of Jeannine Bouiller. Together they went to the roof of his apartment building. Siegmond was setting up the camera when Bouiller gasped and pointed toward the northeast. Siegmond turned and saw the object at about seven thousand feet.

According to Siegmond, the object was dark and elliptically shaped. He set the camera at Infinity and began to take pictures.

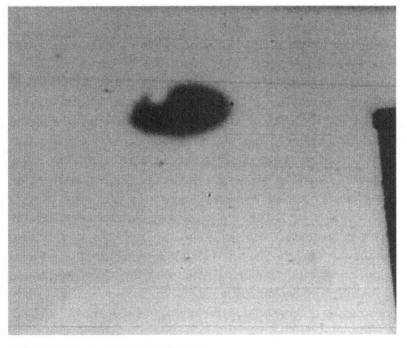

Object seen over New York City

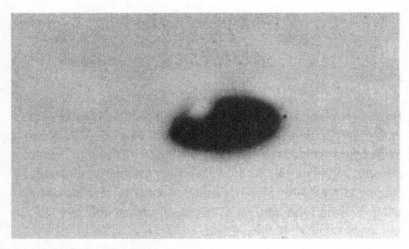

Close-up of the object

He took ten, but only five came out. Siegmond asked the druggist, who had had the pictures developed, where the rest were and was told that was all there were.

The object was in sight for about a minute and a half. It was dark when first spotted but became lighter as it got closer. Siegmond and Bouiller eventually lost sight of it.

The air force rated Siegmond's reliability as dubious at best. The investigating officer noted that Siegmond, after getting the pictures back, took to them to *Life* magazine and to the *New York Journal*, but neither organization wanted to publish them. One of the reporters who looked at the pictures thought they were the result of water vapor on the camera lens.

More important to the air force investigator, however, was the fact that Siegmond "confessed" to an interest in the "Saucer Phenomena" and that he had begun to attend meetings of a local flying-saucer club. The officer noted that "source demonstrated more than a cursory knowledge of the terminology and more notorious names of 'Flying Saucer' literature, e.g. Adamski, Keyhoe, Vaeth."

The air force officer also noted that Siegmond was enjoying the publicity and the speculation caused by his sighting. He wrote, "Considering what might be termed a neurotic susceptibility to give credence to 'Flying Saucers,' Source is considered to be of extremely dubious reliability."

Of course, there was a second witness, Jeannine Bouiller. She didn't say much during the interview and didn't volunteer any new information. They noticed that she would confer with Siegmond before she answered any questions. They did learn that she was employed by the French government's tourist office, but she wouldn't tell them what she did for them.

The investigating officer wrote, "Her description of the sighting coincided identically with that of Mr. Siegmond. In view of her obvious 'rapport' with the opinions of Mr. Siegmond, Source is considered unreliable." This kind of presumption is known, of course, as "guilt by association."

The air force investigators determined, to their satisfaction, that the case was a hoax. One officer noted that the objects in the photographs, that is, a water tank on a roof top, other buildings, and the antenna on top of the Empire State Building, were in sharp, crisp focus. The object, however, was blurred, suggesting that it was closer to the camera than the other objects. If such was the case, then it would mean the object photographed was no more than fifteen feet away, meaning it was a small object.

Nowhere in their findings did the air force suggest that the blurring might be the result of the motion of the object. Clearly, because the other objects were sharp and crisp, the blurring of the object was not a result of camera motion.

One of the officers also noted that, compared with the size of the tower on top of the Empire State Building, the object was more than 750 feet in diameter. He found it impossible to believe that an object that size could fly over a city of more than eight million and no one else would see the object. Of course he didn't allow for the possibility that others did see it but just didn't report it to the air force. Even in 1955 people were worried about the reaction of friends and family when a flying saucer was seen and reported.

But the story didn't end in 1955 with the air force investigation. In 1966 the editors of *Look* magazine produced a special edition about UFOs containing information on dozens of cases including some of the better photographs. Siegmond was again interviewed. He told the same story of photographing Bouiller when she spotted the UFO.

Siegmond told reporters for *Look* that "the Air Force checked the picture and the circumstances thoroughly, and eventually notified me that it was classed among the small percentage of sightings they couldn't explain."

He continued, saying, "It definitely wasn't the planet Venus, or a temperature inversion, or any of those other things that are sometimes mistaken for unidentified flying objects."

Siegmond also said, "I don't know what it was. I was thoroughly checked out on all kinds of aircraft during World War II and I never saw anything like it. I haven't seen anything like it since, and certainly haven't taken any pictures of any."

His statements in *Look,* however, cause us to reexamine the case. Unlike the editors of the magazine, we have access to the air force files, and we know that his statements, eleven years later, simply aren't true. The air force believed that his photographs were the result of a hoax. That would be an explanation, though not one that Siegmond would appreciate.

A series of questions arises: Did Siegmond intentionally mislead the reporter by his statements? Did the air force sugarcoat their explanation in 1955 for public relations? Or did Siegmond actually believe that the air force had found no explanation for his sighting and photographs?

The Heflin Controversy
AUGUST 3, 1965

Location: Orange County, California

Witnesses: Rex Heflin

Craft Type: Daylight disc

Dimensions: Thirty feet in diameter

Primary Color: Gray

Craft Description: Disc-shaped

Sound: None

Exhaust: None

Quality of Photos: Sharp and clear

Number of Photos: 4

Type of Camera: Polaroid

Sources: Project Blue Book files; "The Heflin Case: Then & Now," by Robert J. Kirkpatrick in *The International UFO Reporter,* September 1956; *Scientific Study of Unidentified Flying Objects,* ed. Daniel S. Gillmor

Reliability: 6

Narrative: The pictures taken by Rex Heflin on August 3, 1965, were labeled a hoax by the air force almost immediately. They based their conclusion on their preliminary examination of the pictures made only from copies that had been printed in the newspaper. They believed that the pictures were of a small model, about three feet in diameter, that had been suspended about twenty feet above the ground. Although the air force would engage in another, more lengthy investigation later, they would not alter their original assessment. The case is characterized as a hoax in the Blue Book files.

The story of Heflin's photographs, as it appeared in the Blue Book files, stated:

At approximately 1130 hours PDT on 3 August 1965, Mr Rex E Heflin was driving SSW on Myford Road to check out a matter (not identified) pertaining to his job for the Orange County Road Department. After checking the item he turned his vehicle (Ford "van bus") around and began to head NNE, still on Myford Road. He was still traveling at only a few MPH (three to ten) when he first observed something out of the corner of his eye, out of the left side window of the vehicle. At first he assumed that it was a helicopter until the object was almost directly in front of his vehicle. Because of the unusual shape of the object, Mr Heflin quickly stopped the vehicle and grabbed the Polaroid camera which was on the front seat (SOP for traffic investigators), and took one picture of the object through the windshield. (NOTE: Marine Corps investigators determined that the object was at a bearing of 10 degrees magnetic at the time the photograph was taken.) The UFO continued an ESE course at slow speed. Mr Heflin was able to take two additional photographs of the UFO through the right door window of his vehicle. (NOTE: Marine Corps investigators estimate that the bearing of the UFO in picture nr 2 was about 90 degrees magnetic, and that the bearing of the UFO in picture nr 3 was about 70 degrees magnetic.) This would indicate that picture nr 3 was taken after the UFO had changed course to a NE heading.

Mr Heflin's comment to the Investigating Officer varied somewhat from those given to the Marine Corps investigator. However, it is believed that the undersigned conducted a more thorough investigation and that Mr Heflin's comments which are noted below are essentially correct. Mr Heflin stated that he had attempted to use his two-way radio once or twice just before he sighted the UFO and could neither transmit nor receive any signal although the radio panel lights indicated that the radio was operational. Detailed questioning indicated that this definitely occurred before the UFO sighting and not during the UFO sighting. Mr Heflin stated that after the UFO disappeared, he attempted to use his radio and found that it was working normally. He also stated that he had not had any other radio malfunctions in the preceding weeks.

Just after taking the third picture of the UFO, Mr Heflin heard a vehicle approaching from the rear. Concerned that he might have parked in an awkward position, he turned around to see if there was enough road clearance for the vehicle to pass him. Noting that he was on the shoulder of the road, he immediately turned again to look at the UFO but found that it had "disappeared into the haze."

The Investigating Officer has noted some skepticism on the part of several individuals as to whether or not Mr Heflin could have observed the UFO, stopped his vehicle and taken three photographs—all in a 15 to 25 second time period. A check was made with people who are familiar with this particular model camera, and it was determined that an experienced man could easily take three photographs within a 12-second time period.

At the end of the working day, Mr Heflin returned to his office in the Road Department building and showed his photographs to several co-workers. They were subsequently filed in a desk drawer. After a few days (and after several duplicate photographs, conversations, comments, and general bull sessions), a friend of Mr Heflin who is a former employee of the Road Department convinced Mr Heflin that they should try to sell the photographs to LIFE Magazine. The friend was unidentified, however Mr Heflin's supervisor, Mr Kimmel, substantiated this bit of the story. The friend called the LIFE Magazine office in Chicago and was informed that they were interested. Subsequently, the friend mailed the pictures to the Los Angeles office and presumably the photographs were forwarded to the main office in New York (or Chicago?).

About two weeks later the photographs were returned from New York (?) directly to Mr Heflin without written comment. At about the same time the Los Angeles office telephoned Mr Heflin to say that the main office had declined to utilize the pictures "because it was too controversial at the time". Time passed and apparently more copies of the pictures were made and handed out to various friends and friends of friends, until most of Santa Ana was saturated with the UFO pictures.

One of these pictures was obtained by a druggist who then apparently showed it to a friend, a customer who worked for the Santa Ana Register. On or about 18 September, Frank Hall of the Santa Ana Register contacted Mr Heflin, borrowed the three original prints, returned the originals to Mr Heflin and wrote an article which was published with the UFO picture (Nr 1) in the Santa Ana Register on 20 September 1965. Two of the three photographs were released by the Register on 20 or 21 September to UPI. One or both of these pictures and accompanying articles were published by various newspapers on 21 September 1965.

There are additional points to be made here. Heflin apparently loaned the original prints to anyone who asked for copies of them. First he gave them to the marine officer who investigated the case and did not ask him for any sort of a receipt. The marine quickly returned the originals. Then, according to Heflin, a man claiming to be an officer assigned to the North American Air Defense Command (NORAD) contacted Heflin. He was given the originals, again without Heflin asking for a receipt.

The air force tried to find out who the NORAD officer was but had no luck in doing so. Others noted that NORAD wouldn't be investigating UFOs and certainly wouldn't have taken the originals without making arrangements to return them. Air force investigators suggested that because NORAD would not be routinely investigating, and because there was no evidence that such an investigation by NORAD had taken place, Heflin's story was a hoax. They didn't seem to consider the possibility that the man claiming to be from NORAD was an impostor.

Heflin's chronology and reports were also disputed. Air force records suggest that the air force only became interested in the case after the public release of the photographs. Yet, in the Project Blue Book files, there is a report of the photo analysis of the pictures dated August 14, suggesting an air force investigation within two weeks of the event and before the public interest.

The Condon Committee asked the air force about the photo analysis and was told that the report's date was a typographical error. Air force officers had noticed the problem earlier and had asked questions about it. No one explained why the date hadn't been corrected on the form when the error was noticed or why a notation about the inaccurate date hadn't been attached to the report.

With the question of the date answered to the satisfaction of the Condon Committee investigators, they began to pick apart the rest of Heflin's story based on exactly that same sort of error. For example, Hartmann noted that in his original testimony, Heflin had spoken about a single visit from a single NORAD officer. According to a September 25, 1965, report filed with NICAP Heflin had said that "a man with a briefcase later called . . . and said he was . . . and that he would like to see . . . [Heflin] agreed to loan the pictures to him providing he would . . ."

Colorado investigators interviewed Heflin on January 15, 1968, and learned Heflin now believed there were two officers. The

Condon investigators asked, "Why is it that you are now clear on there having been two NORAD visitors, while on the very next day the Air Force man came away with the idea that a man came up and flashed his card . . . ?"

According to Hartmann in the Condon Committee report, "He [Heflin] immediately replied in effect that only one man showed his card. He repeated that there were two men, in their early thirties, but that one stood back while the other did most of the talking. Since two independent reports from the next three days clearly indicate one visitor, while the witness has since insisted there were two, the 'NORAD episode' is still regarded as open to serious question."

But this discrepancy isn't a very important issue. It can easily be explained as a simple mistake by Heflin. Human memory is a very complex thing and there are changes and mutations in memory each time it is accessed. The importance attached to the number of men who appeared at Heflin's home is the same as the importance attached to the date of the air force photo analysis. It is a mistake that is insignificant and irrelevant.

Analysis of the photographs should have taken primary importance in the investigation, but it was the testimony of Heflin that was scrutinized in detail. There were, however, some analyses of the pictures that could have been significant.

First, is the misdated air force report, numbered 65-48. In paragraph 2, under the heading "Analysis," it was reported that "although it is not possible to disprove the size of the object from the camera information submitted and distances to the object quoted in the report by Mr. Rex E. Heflin, we feel that the following is the true case. The camera was probably focused on a set distance and not on infinity, as the terrain background was blurred on all three photographs. The center white stripe on the road and the object appeared to have the same sharp image. Therefore, it is felt that the object was on the same plane as the center white stripe or closer to the camera and could not possibly be the size quoted in the report. Using the width of the road as a factor, it was estimated the size of

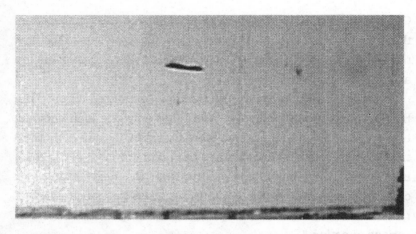

Air Force attempts to duplicate the Heflin photographs

the object to be approximately one to three feet in diameter and approximately fifteen to twenty feet above the ground."

The air force investigators also reported, "A test was conducted by the FTD Photo Analyst and Photo Processing personnel with the results showed on the attached photos. The photographs were taken with a Polaroid Camera, Model 110A using 200 ASA film. . . .

156

The object seen in the photographs was a 9" in diameter vaporizing tray, tossed in the air approximately 8 to 12 feet high at a distance from the camera of approximately 15 to 20 feet. The result of the test shows surprising similarities between the object on the test photography and the object on Mr. Heflin's photography."

On the other side of the coin, however, was the photographic analysis conducted by Ralph Rankow, a NICAP member, who said that his preliminary findings supported the authenticity of the pictures.

In all fairness it must be noted that no one expected the air force to find anything other than the pictures were some kind of fake. And no one expected NICAP to find anything other than the pictures were authentic. The political agendas of the two organizations are apparent in their analysis of Heflin's pictures.

There was also a controversy surrounding the fourth of the Heflin photographs. When he returned to the office on August 3, he showed the three pictures of the UFO to his coworkers. He allowed friends to have the pictures copied. He showed those pictures to investigators and discussed them with reporters. Hundreds were aware that he had taken three pictures of the UFO.

What no one knew at the time was that Heflin had taken a fourth picture of a smoke ring that was allegedly left by the UFO. It was taken from farther down the road and could be an important clue as to the authenticity of the case.

The documentation available, from the air force file, the NICAP investigation, and that of the *Santa Ana Register,* makes no mention of this picture. The Condon Committee investigators wrote, "During the early NICAP interview the presence of a fourth photo was not recorded, although the ring was apparently mentioned. During the Air Force interview, the witness not only did not mention the smoke ring or fourth photo, but gave a somewhat different description of the disappearance of the UFO."

The air force report noted, "Just after taking the third picture of the UFO, Mr Heflin heard a vehicle approaching from the rear. Concerned that he might have parked in an awkward position, he turned around to see if there was enough road clearance for the vehicle to pass him. Noting that he was on the shoulder of the road, he immediately turned again to look at the UFO but found that it had 'disappeared into the haze.'"

Heflin, when questioned about this part of his account, said that he had been advised by a NICAP investigator to withhold some of the information from the air force. An attempt by Condon Committee investigators to clear up the problem resulted in Heflin denying that he had invented any of the testimony. Everything he had said he claimed was true, but that didn't explain why there was but a single reference to the truck diverting his attention. The NICAP investigator thought that this one aspect of the tale might be a falsehood.

The problem with that thinking is that once you have caught the witness in a deliberate lie, it calls all other testimony of that witness into question. If he would lie about one aspect, why not others, and how do you know when he is lying and when he is telling the truth?

The lie seemed to be designed to explain how the photograph of the smoke ring was taken. It was designed to explain an apparent discrepancy in the time line of the sighting event. The appearance of the truck that diverted Heflin's attention covered the point easily. But if this particular of the story was true, why didn't he mention it again, after telling the air force about it?

At any rate, these changes in the story were enough for the skeptics and debunkers to dismiss the Heflin tale as a hoax. To them, these "internal inconsistencies" proved that Heflin had invented the story for the publicity and excitement.

Those who wished to dismiss the Heflin story also directed attention to the clouds that appeared in the fourth picture. According to weather records from five different reporting facilities, there were no clouds in the sky on August 3. If true, that would suggest that picture number four was not made at the same time as the others, and that fact would throw off the chronology.

Nearly twenty years after the Condon Committee report was issued, Heflin would tell Robert J. Kirkpatrick, "[I] lost considerable credence in the scientific community when I saw the weight they placed on a review of the weather data supplied by five agencies in the area. Anyone familiar with the weather in Orange and Los Angeles counties knows what extreme variations exist in weather conditions from one locality to another and how quickly conditions can change."

However, Heflin claimed that there were additional witnesses to the UFO appearance. He told Kirkpatrick that a private pilot who had been flying from the Fullerton, California, airport also reported the UFO. The pilot never came forward to corroborate the sighting. Heflin also told Kirkpatrick, "In fact, a member of another field crew said he saw the same UFO. But he died in a traffic accident before any investigators could interview him."

These sorts of witnesses are of no value now because we have no record of their statements or who they are. A man who died before he corroborated the story, who left no legacy of his report, does nothing to validate the Heflin case. There is no way to learn if he actually saw the UFO, if he suggested he had because he wanted some of the spotlight, or if this was just Heflin's attempt to show that others had, in fact, seen something.

The Condon Committee investigators drew no real conclusion in the Heflin case. They wrote, "From the point of view of the Colorado study the principal question of concern is: Does a case have probative value in establishing the reality of unusual aircraft? In a case like this, where both the observer and the photographs clearly allege an extraordinary vehicle, a second question is, of course, automatically implied: Does the case represent a fabrication or was the object a true unknown? . . . We are concerned only with establishing evidence as to whether or not there exist extraordinary flying objects.

"In that context, this case is equivocal."

In the committee report's conclusion section, it was noted that "the strongest arguments against the case are the clouds in photo four and the inconsistent early records regarding the 'NORAD' visitors. The photos themselves contain no geometric or physical data that permit a determination of distance or size independent of the testimony. Thus the witness's claims are the essential ingredients in the case. The case must remain inconclusive."

That wouldn't be the end of it, of course. The *Orange County Register* published an article by Amy Wilson on July 22, 1997. According to the article, Ed Riddle, a technical writer in Menlo Park, said that he had first seen Heflin's pictures when a man brought them into the lunchroom. Riddle said that the man said that a friend or neighbor had "rigged up a toy train wheel and some monofilament fishing line, hung them out his truck window, shot them and would, maybe, just take them to the paper for some fun."

According to Riddle, in 1965, he called the *Orange County Register* to let them know that the pictures were a joke. But his tale was met with "gruff talk about how if a man didn't have any proof, they didn't want to hear about it."

Amy Wilson also reported that a second caller had told the *Register* that Heflin lost his job because of the controversy over the pictures. That wasn't true either. Heflin worked at his job for thirty years. In fact, he had retired in 1978 as Chief of the Traffic Investigation Division after a disabling injury. Clearly Heflin's career was not adversely affected by the UFO photographs.

There is probably one other comment that should be made. Many inside the UFO community worry about how much money is made by those who claim to have seen something strange. Heflin's pictures have been used in dozens of magazine articles and books about UFOs. When he took the pictures in 1965, he didn't bother to have them copyrighted. Anyone could use them without having to pay for the privilege. Heflin never received any money for the use of his pictures.

Multiple or Single Witness?:

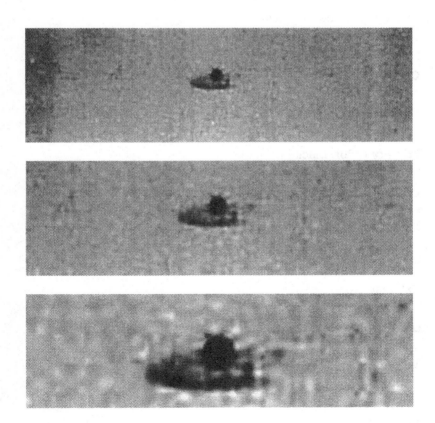

Location: Wall Township, New Jersey

Witnesses: Robert J. Salvo
                     Diane Salvo (possibly)

Craft Type: Disc

Dimensions: About fifty feet in diameter

Primary Color: Black

Craft Description: "Looked as big as a jet liner, saucer shaped, had a revolving conning tower in the center."

Sound: Humming, whistling

Exhaust: None

Quality of Photos: Sharp and clear

Number of Photos: 5

Type of Camera: Pala, with only two settings at fifteen feet and Infinity.

Sources: Project Blue Book files; *Look* "Flying Saucers," special one-time issue, 1966); *Asbury Park Evening Gazette* (New Jersey) May 13, 1966; *The Daily Register* (New Jersey) May 12, 1966; *Long Branch Daily Record* (New Jersey) May 12, 1966; *Newark Evening News* (New Jersey) May 12, 1966

Reliability: 6

Narrative: Just before his family was scheduled to move, Robert Salvo decided that he wanted some pictures of his dog near the old gravel pit. He heard a loud whistling and humming sound and looked up. At first he thought it was some kind of aircraft, then some sort of joke played by his friends, and finally that it was none of those things. Salvo took five photographs as the craft hovered over him for about ten minutes. Finally it shot back up into the clouds and disappeared.

When he returned home, he told his mother about the sighting. She said that if he was sincere about the sighting, he should send the pictures and a letter to the local Civil Defense authorities, which he did.

Frank L. Wilgus, a Civil Defense coordinator, reported the sighting to army authorities at Fort Monmouth, New Jersey, on April 20, 1966, the day after he received the information from Salvo. The army security personnel did nothing about it because the sighting was now about three weeks old. Officials contacted by newspapers in May stated that had the information come in sooner, they would have sent it to their headquarters, which would have passed it on to the air force. The army, after all, was not tasked with the investigation of UFOs.

Eventually the sighting and the photographs reached Project

Blue Book. The air force attempted to learn more about the incident and contacted both Salvo and Wilgus. However, when they tried to analyze the photographs, they ran into trouble. They needed some specific information, but that information was just unavailable.

Major (later Lieutenant Colonel) Hector Quintanilla wrote to Wilgus asking for specific information about Salvo's camera. He needed the focal length and the camera type so that they could determine information about the lens used.

Wilgus wrote back with the information, which the air force found to be inadequate. The camera was cheap. It had cost only eighty-eight cents and was known as a Pala. It had two focal lengths, one for objects closer than fifteen feet and one for objects farther away than fifteen feet. According to Wilgus, Salvo's camera used 120 film, and the film happened to be made by Kodak.

Analysis of the camera was attempted by the air force. Charles A. Burton did the work. Burton, in his report of July 28, 1966, wrote, "Due to having only the focus position settings available, a check was made of local sources to determine the focal length. This revealed only that the camera was a cheap make, not for sale on the open market, and primarily made for premium distributions. A camera of this type would normally use standardized parts and probably have a focal length of approximately 3 inches."

With that question of camera specs settled, at least partially, Burton began his calculations concerning the photos. He wrote, "Assuming an approximate 3 inch focal length, several sizes were computed for selected ranges between 10 feet and 1 mile. The only range available from the observer is found on the back of one of the original prints. . . . The comment is as follows, '. . . I'd say it was about a quarter of a mile up.'"

Burton calculated that "the range of 1320 feet reflects this [above-mentioned] estimate. Photograph #2 was used for these calculations as it is the only one with an estimated range available. The results were as follows:

"Observer estimated range—1320 feet—diameter 52.8 feet

at range of—10 feet—diameter .4 feet

at range of—15 feet—diameter .6 feet

at range of—20 feet—diameter .8 feet

at range of—50 feet—diameter 2 feet

at range of—100 feet—diameter 4 feet

at range of—5280 feet—diameter 211 feet."

Burton concluded his report, writing, "Further analysis would require reliable estimates of distance from camera to object (range) for each photograph and exact focal length of the camera."

Because of the imprecise nature of the information about the camera, and the lack of estimates for the range on each of the photographs, the air force concluded that the case had "insufficient data for a scientific analysis."

There are a couple of other points to be made. First, had the air force investigated the case thoroughly, they could have gotten the precise information they required for their analysis. Certainly, rather than writing letters to the Civil Defense coordinator, they could have interviewed Salvo in person, examined the camera, and gotten the range information they needed.

Second, it is possible that the sighting was witnessed by others. There is a confusion in the file as to whether or not Salvo was alone when he took the pictures. In one newspaper account, Salvo's older sister, Diane, said that she had seen an object shortly before Robert's sighting, which was flashing red, white, and blue lights. The neighbors also saw this. It is unclear if this sighting was made days before, or only minutes before the photographs were taken.

In another newspaper article, it mentioned that Salvo's sister was with him and both were anxious for someone to explain what it was they had seen. If Salvo's sister was, in fact, with him, it changes the nature of the case.

Another Teenager with a Photograph:

Location: Milledgeville, Georgia

Craft Type: Domed disc

Dimensions: Fifty feet in diameter

Primary Color: Gray

Sound: None

Exhaust: None

Quality of Photos: Sharp and clear

Number of Photos: 2

Type of Camera: Polaroid Swinger

Sources: Project Blue Book files

Reliability: 1

Narrative: There had been sightings in the area of Milledge-ville, Georgia, for several days, as noted by Eugene A. Ellis, the chief of the Milledgeville police. He wrote to Lieutenant Colonel Hector Quintanilla at Wright-Patterson Air Force Base. Witnesses to nonphotographed events included four police officers, two deputy sheriffs, and a couple of tourists. None of these people were witnesses to the sighting in which pictures were taken.

When the Georgia sightings took place, the University of Colorado study was in full swing. The witness who had taken photographs of his sighting was interviewed by a number of investigators from Colorado, including Dr. Roy Craig and John Ahrens, as well as a number of air force officers.

They took the witness back to the scene so that he could describe for them exactly what he had seen and so that the investigators could measure angles and distances. They recorded their interview with the witness.

According to the transcript, the witness was a boy who had told his mother that he was going into the woods to hunt for UFOs. He took his BB gun and his Polaroid Swinger camera with him. He had walked to one of his favorite places and was passing time, kicking at the rocks and climbing trees when he looked up and saw the object. He thought that it was hovering but then noticed that it was moving slightly.

After watching it for about thirty seconds, he decided to take a picture of it. Inexplicably, he sat down on a small bush, pointed his camera and took the photograph. He told investigators, "The pictures I took, I pulled the first one out and looked at it and set it down. Then I pulled off the second one and looked at it, and set it down. And then I heard my friend coming—someone calling for me—so I grabbed up the pictures and ran up there like—I was so scared that I was yelling like something was after me or something. I showed him [the friend] the pictures, and he looked at

168

them for five or ten seconds, and I kept pulling him toward the road, and when we got down here, it [the UFO] was gone. It started to leave, I don't know exactly if it started to leave, but it started going higher up in the air when he [the friend] started calling—about a little before he started calling—it started rising in the air. It started picking up speed, it looked like."

During the questioning, it became clear that the witness, along with three of his friends, had formed a "flying saucer club." They discussed the topic and spent time searching for UFOs. The witness also mentioned that he was interested in science fiction. Neither activity is uncommon in teenage boys.

The air force wrote the case off as a probable hoax and declared that there was insufficient data for a scientific analysis of the pictures because the originals had not been submitted. They noted that the boy was the only one to see the object and that he had gone out with the express purpose of finding a flying saucer.

It seems clear that the interest in UFOs in the local area, and the number of sightings there that had been reported to the police, might have suggested to the boy the idea of taking a picture. It is interesting that the air force, in communication with the chief of police, made no effort to investigate the other sightings. Instead, an investigator was sent from the University of Colorado to interview the teenager.

Although that fact makes some sense on one level, it doesn't on another. Yes, the boy had pictures to support his story, but there were no other witnesses. Four police officers and two sheriff's deputies who had seen UFOs in the days before the photographs had been taken might have provided some interesting information. They, however, were not interviewed by either the Colorado investigators or by air force officers who had escorted the investigators into the area.

The air force conclusion about the case, that it was a hoax is probably accurate here. The boy's actions, as described by him to the investigators, do not make sense. That, in and of itself, suggests a hoax. And, without the corroboration of another witness who could independently verify his information, there is no reason to draw a different conclusion. This is one of those times when the air force got it right.

One of the First Triangles:

SEPTEMBER 3, 1967

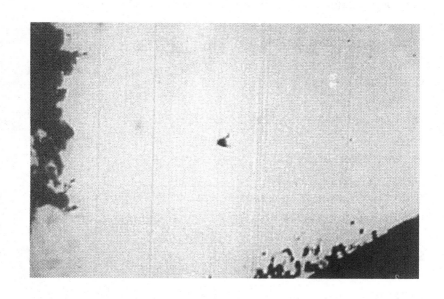

Location: Plaquemine, Louisiana

Witnesses: Charles W. Pierce

Craft Type: Triangular

Dimensions: Two hundred feet long

Primary Color: Red

Craft Description: It appeared to look like the top of a container that had been heated red-hot, and it was spinning.

Sound: None

Exhaust: None

Quality of Photos: Sharp and clear

Number of Photos: 1

Type of Camera: Polaroid Swinger

Sources: Project Blue Book files

Reliability: 4

Narrative: The witness, Charles Pierce, said that he had spotted a spinning, red object that was heading in a northwesterly direction. With his Polaroid camera, he took a single photograph. He then communicated with the air force, sending the picture on to them.

The air force labeled the case as a possible hoax, noting that, in his description of the sighting, "the observer described almost perfectly the object that appears in his photo. However, photo analysts indicated that the image was a probable processing blemish."

The air force stated further that the analysis "showed the unidentified image to be generally triangular in shape and positioned in the approximate center of the frame format. It was noted that the unidentified image, a portion of a building roof and foliage in the foreground exhibited the same amount of lateral image displacement. This condition resulted from camera movement/vibration recorded by the relatively slow shutter speed of the Polaroid Swinger camera. The source stated that the object had a 'fast forward speed.' If this statement were true, the unidentified image would have exhibited blurring which resulted from a shutter speed not fast enough to stop the motion. The photography indicated the camera, and the objects recorded, were motionless except for the small camera movement/vibration induced image displacement when the exposure was made. Examination of the photography by the Chief of Quality Control, indicated the unidentified image to be a probably processing blemish."

Advertising UFOs:

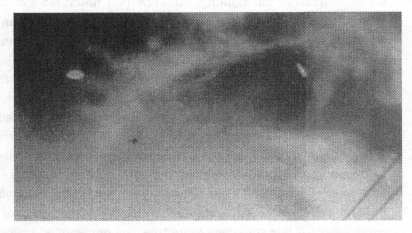

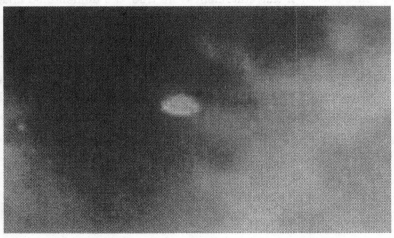

Location: Flushing Queens, New York

Witnesses: Not identified

Craft Type: Domed disc

Dimensions: Thirty feet in diameter

Primary Color: Reddish, reflecting the setting sun

Craft Description: Domed disc with indentations on the dome

Sound: None

Exhaust: None

Quality of Photos: Sharp and clear, color

Number of Photos: 3

Type of Camera: Nikon Nikinos

Sources: Project Blue Book files

Reliability: 4

Narrative: The witness, a president of an advertising agency, told air force investigators that he and others were traveling from New York to Washington, D.C. Because the sky was beautiful, with the sun setting, he decided to take several pictures, holding the camera outside the car window, as they traveled at sixty-five miles an hour. The witness told the air force that his vision was limited because they were in a sports car, and that he didn't see anything strange in the sky. The objects were seen when the slide film was developed.

After they returned to New York and the film was projected for the first time, the witness and his friends saw the objects near a bridge. The witness had the pictures blown up so that the objects were about a foot in diameter. He said that they were domed discs, with indentations on the dome. He believed they were metallic and that they were reflecting the light of the sun. According to the witness, based on his observation of their position at the bridge, the objects were moving to the south, following the river.

The witness knew that the air force had been investigating UFOs and thought they would be best qualified to analyze the photographs. Air force officer Lieutenant Conaway, from the Information Office at Suffolk County Air Force Base (reporting to Lieutenant Colonel Hector Quintanilla of Project Blue Book), investigated. He assured the witness that his original photographs would be returned but that the air force couldn't properly analyze anything other than the original negatives. Since these were color slides, the air force officer wanted the original transparencies.

Conaway was concerned because the man told him that the photographs were valuable. According to the report, in a sentence that was underlined, Conaway noted that the man had said he "had numerous money offers from magazines."

Conaway was told by Quintanilla that air force regulations demanded that he send the original negatives and that the forms be completed properly. Quintanilla then told Conaway that the witness was probably trying to get the "Air Force to say that his photographs are authentic. Well, all photographs were authentic, but UFOs aren't."

Although the photographs were provided for the air force, apparently the paperwork, that is, the report by the witness, was not completed quickly enough. The air force returned the photographs before they received the report. Therefore, according to the air force, the case was labeled as having "insufficient data for a scientific analysis." In this case, it meant that the witness had not complied with air force requests to complete their rather lengthy forms.

One of the air force forms, in which an officer was asked specific questions, ended with a request for a summary. It directed the investigator to "state your own personal evaluation of the report. What do you think the object was? Do you think something other than the sighting motivated the caller? Include anything which may add to the objectivity of the report. Include your evaluation of the caller's reliability."

Concerning his interview with the witness, Sergeant Robert Becker filled out the form and wrote, "According to the caller's description, he did photograph some type of object, rather than an optical illusion. I would not however, exclude the possibility of uncommonly shaped high or middle clouds. I did not form any opinion of some motivation for calling. I did not[e] one apparent contradiction; he said he was just photographing a beutiful [sic] sunset, yet his discription [sic] of the photos sounds to me like he might have, in fact, been shooting at the objects."

The problem here is that the witness had, quite clearly, studied the photographs for a long period before alerting the air force. He told air force investigators that those who had traveled with him had also studied the photographs. Their study certainly could have contributed to their telling of the story, which suggests that the witness had actually tried to photograph the objects rather than just a beautiful sky.

The air force attitude here is also of interest. Quintanilla's bias, that photographs are real, but UFOs are not, is interesting. He suggests that, by this point, June 1968, the air force was just attempting to explain rather than investigate.

A Diamond-Shaped Craft:

Location: Miami, Florida

Witnesses: Not identified

Craft Type: Diamond

Dimensions: Thirty feet in diameter

Primary Color: White

Craft Description: Diamond shape surrounded by halo

Sound: None

Exhaust: None

Quality of Photos: Sharp and clear

Number of Photos: 2

Type of Camera: Nikon F with 50 mm lens

Sources: Project Blue Book files

Reliability: 5

Narrative: In 1968 two reporters for the *Miami News* visited the local air force base with copies of photographs that had been taken on the evening of September 10 during a thunderstorm. The object was not seen by the witness. He was leaving the lens open to photograph the storm. Only the last two frames showed the object.

First Lieutenant Robert R. Cloar, an assistant information officer and the base UFO project officer, talked to the photographer. In his report to Wright-Patterson Air Force Base, he wrote, "He [the photographer] has been a professional photographer for 15 years. . . . The News is a reputable paper with the second largest circulation in the Greater Miami area. . . . [the photographer] has no formal education or training nor does he have any pilot experience. This is not pertinent to the case because he never saw the object. I believe . . . [his account] to be reliable but a little UFO controversy never hurts a photographer or the paper's circulation."

On November 20, 1968, Lieutenant Colonel Hector Quintanilla wrote to Lieutenant Cloar, "The photograph that you forwarded in your letter of 11 September 1968 is being returned without an analysis being performed. In your telephone conversation with Lt. Marano [of the Project Blue Book staff] of this office you indicated that you would attempt to obtain the original negatives. To date we have not received the negatives. A complete analysis cannot be made without the original negatives."

With that, the air force washed its hands of the case. It was listed as having "insufficient data for a scientific analysis." Once again, there was no explanation for a sighting, but the object reported in the case was kept from being labeled as "unidentified."

Another Teenager's Photographs:

OCTOBER 13, 1968

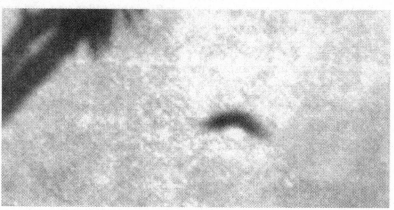

Location: North Forestville, Maryland

Witnesses: Fifteen-year-old boy

Craft Type: Daylight disc

Dimensions: Twenty-five feet in diameter

Primary Color: Silver

Sound: None

Exhaust: None

Quality of Photos: Sharp and clear

Number of Photos: 1

Type of Camera: Polaroid

Sources: Project Blue Book files

Reliability: 1

Narrative: The official air force investigation of the North Forest-ville case revealed that a photograph of a sighting was taken by a fifteen-year-old student. He claimed that the object was in sight for about ten or fifteen minutes. There were no other witnesses to the sighting, no radar corroboration, and only the single photograph.

The air force noted that "There was also some giggling in the background as the initial report was given over the telephone. Although the observer would not release the original photo, the object looks very similar to a small suspended model or similar object."

This is one case where the air force analysis seemed to be accurate. There was no reason to reject their opinion.

Part IV:

THE INTERNATIONAL SITUATION

As has been said time and again, the UFO phenomenon is a world-wide problem. Sightings have been made in every country in the world, and the major UFO organizations have often boosted the number of their representatives in the vast majority of the world's countries. In fact, the exceptions to this trend, until recently, had been some of the nations behind the Iron Curtain and a few of the Third World countries where there simply were no communications capabilities.

Today, with fax machines, the Internet, satellite communications, computers, and air travel that puts, within a very short time, the majority of the world in contact with just about any point on the globe, the UFO situation is more evident. A UFO sighting in North America or Africa or Europe might be reported to those interested in other countries almost before that object has disappeared. There is no question about the international nature of the UFO phenomenon today. But that wasn't always the case.

In the past, it seemed that UFO phenomena were centered over the United States. Although sightings made elsewhere were sometimes reported in the United States, often they were simply ignored. That is why those first becoming interested in UFOs would often ask if there were UFOs spotted in foreign nations.

The real point is that many of the aspects of the UFO phenomenon that we take for granted were first reported in a foreign nation. Only after someone in the United States made a similar report did we realize exactly what was happening regarding sightings.

Although the first real wave in which the flying saucers were reported, as opposed to the tales of the Great Airship, the ghost rockets, or the Foo Fighters, was in the United States in 1947, there were many sightings from other parts of the world that same year. Sometimes those reports found their way into the American newspapers with a sort of "the-rest-of-the-world-is-now-seeing-these-things-too" attitude, but more often they were simply ignored.

Take, as an example, the reports of creatures from inside UFOs. While there were a few such reports inside the United States in 1947, there were also reports from South America, some of them coming even before Kenneth Arnold made his sighting in June 1947. In February of that year, a witness, identified in the UFO literature only as C.A.V. because he wished to remain anonymous, spotted a disc-shaped object hovering about six or seven feet off the ground. He stopped and got out of his car, thinking that he would walk over to see what it was. The distance was farther than he had thought.

When he reached the object, which he described as sand-colored and very shiny, three figures came out of it. C.A.V. saw no door or hatch, but the creatures seemed to have come from the inside. He described them as looking like mummies, a description that would surface again in other reports, that they had sand-colored skin and had legs that were fused together.

Later, as tales of ships were being told in the United States, more reports of creatures were coming from Europe. In August 1947, as the interest in flying saucers was on the wane, Professor R. L. Johannis reported his encounter with alien beings. He said that he was in the mountains in the extreme northeastern part of Italy when he saw, on a rocky river bank, a large red object that seemed to be shaped like a lens. He couldn't see it well and put on his glasses. It was clear to him that the craft was something unusual.

Near the object he spotted a number of alien beings. Johannis and the aliens began to approach one another, but when they were a few paces apart, they stopped. Johannis suddenly seemed to have no strength. Although paralyzed, he could still observe the details. Thinking about it later, he believed that he was standing near one creature for only two or three minutes.

Eventually one of the creatures raised a right hand to its belt. There was a puff of smoke, or a ray of some kind, and Johannis fell to the ground. The pick that he had held was yanked from his hand, as if grabbed by some invisible force.

Two creatures walked closer and picked up the pick. Johannis made a "fantastic" effort and managed to sit up. As he braced himself with his hands so that he didn't fall, the two creatures walked back to the craft. They climbed inside and it then shot into the air.

In the United States, there were few tales of alien beings. Those who told such stories were normally ignored. It was, to some, perfectly normal to see something strange in the sky; it was fine to see the craft close to the ground, or even landed; but it was not proper to see the flight crew from the interior of the craft. In other parts of the world, however, alien-being stories were being told much more frequently. In 1954 there was a major wave of creature sightings in Europe, specifically France, and throughout South America.

Typical of those sightings was one of hairy dwarves seen by two men, Gustave Gonzales and Jose Ponce in Caracas, Venezuela. While driving to Petare, the two witnesses spotted a glowing sphere-shaped object about ten feet in diameter that nearly blocked the road. Gonzales stopped the car and the two men got out to investigate. A small, hairy man approached, and Gonzales grabbed him, hoping to take him to the police. The creature fought back, shoving Gonzales and hurling him away.

Ponce, frightened by the encounter, ran toward the police station only a couple of blocks away. As he ran, he noticed another two of the little creatures, both carrying rocks and chunks of dirt as they jumped into the sphere.

Now the first little creature, with its claws out, attacked Gonzales. Jerking out a knife, Gonzales fought back, stabbing at the alien. The knife struck the shoulder but slipped off, as if the creature was made of solid metal.

A fourth creature stepped from the craft and shot a beam of light at Gonzales. Momentarily blinded, he stumbled back away from the first alien. All the creatures then retreated to the craft, which shot into the air after they boarded.

Gonzales ran to the police station where both he and Ponce were suspected of being drunk. An examination revealed scratches on Gonzales. A couple of days later a doctor confirmed that he had seen the fight but hadn't remained on the scene. He didn't want to get involved.

South America also led the way into reporting abduction phenomena that would become so prevalent in the 1990s. Prior to the alien abduction of Antonio Villas-Boas in Brazil in 1957, no one had

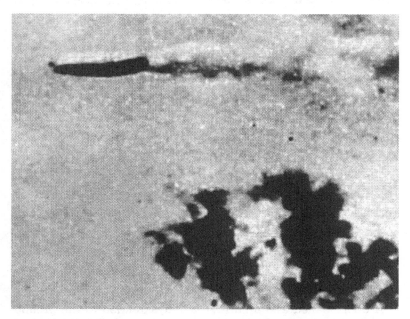

A Peruvian farmer took this photograph over the jungles of the Madre de Dios area in Peru in 1952.

ever suggested that alien beings were interested in the biology, physiology, genetics, or reproductive systems of humans.

The case was first reported on February 22, 1958, by a Brazilian farmer named Antonio Villas-Boas. He was interviewed by the Aerial Phenomena Research Organization (APRO) representative in Brazil, Dr. Olavo Fontes, and a newspaper columnist Joao Martins.

According to the report, on October 15, 1957, Villas-Boas was working in the fields late at night. At one in the morning (October 16, 1957), he noticed an extremely bright red star overhead. As he watched it, he realized it was moving, growing larger as it approached him. He hesitated, wondering what to do, and in those few seconds, the light changed into an egg-shaped craft descending toward his freshly plowed field. It came to a hover over him, its light so bright that Villas-Boas could easily see the fields around him.

The light seemed to come from the front of the craft, which looked like an elongated egg. There were purple lights near the large red one, and there was a small red-light on a flattened cupola that spun rapidly. As the object slipped toward the ground, three telescoping legs slid out from under it.

Villas-Boas realized that the legs, like those of a camera tripod, were for landing. It was now time to escape. He turned his tractor and stopped on the accelerator. Before he had driven far, the engine sputtered and died and the headlights faded out. He tried to start it, but the ignition didn't work. He then opened the door on the side away from the alien craft, and tried to run.

He had taken only a few steps when something touched his arm. He spun and faced a short creature. Struggling to escape, he put a hand on the creature's chest and pushed. The alien being stumbled back and fell, but three other creatures grabbed Villas-Boas and lifted him up. He twisted, kicked, and jerked his arms. He shouted for help, but the beings held him tightly, moving slowly toward the craft.

A door opened and a narrow ladder extended to the ground. The aliens tried to lift him into the ship, but he grabbed at the narrow railing, holding on to it. One of the creatures peeled his fingers from the flexible metal, and he was forced upward, through the door.

For several minutes, Villas-Boas and the aliens stood in a room inside the craft. The creatures talked among themselves in a series of low, growling sounds while they held on to Villas-Boas. Eventually they began to strip his clothes from him carefully so that they didn't tear anything. When he was naked, one of the aliens began to "wash" him with an oily-looking liquid that made him shiver as it dried.

Villas-Boas provided a description of the alien beings. While it is detailed as to what they wore, it provided few clues as to what they actually looked like. He indicated that they were all small, no more than five feet tall. He said, during his long interview with Dr. Fontes, "I must declare that up to that moment I hadn't the slightest idea as to how those weird men looked nor what their features were like. All five of them wore a very tight-fitting siren-suit, made of soft, thick, unevenly striped gray material."

He was forced deeper into the ship, into another, smaller room where blood samples were taken from under his chin by two figures holding two rubber-looking pipes. After the samples were taken, he was left alone for an hour or more. Then, feeling tired, he sat down on a large couch and noticed a strange odor in the room. From the walls, about head level, he noticed a gray smoke pouring into the room. Its thick, oily odor made him physically ill. He fought the feeling for several minutes but finally vomited.

Feeling better, Villas-Boas sat back on the couch and waited. Eventually there was a noise at the door, and he turned to see a woman entering. Like Villas-Boas, she was completely naked. She moved slowly, walked toward him, and embraced him, rubbing herself against his body.

The woman was short, reaching only up to his chin. She had light, almost white, hair that looked as if it had been bleached heavily. The woman had blue eyes that were slanted to give her an Arabian look. Her face was wide with high cheekbones but her chin was very pointed, giving her whole face an angular look. She

191

was slim, with high, very pointed breasts. Her stomach was flat and her thighs were large. Her hands were small, but looked like normal human hands.

When the door closed, the woman began to caress him, showing him exactly what she wanted. Given the circumstances, Villas-Boas was surprised that he could respond at all, and then felt sexually excited. Later he would suggest that his state of arousal was induced, maybe by the liquid that had been rubbed on him.

She kept rubbing him, caressing his body, and within moments they were together on the couch. Villas-Boas responded to her, and before he realized what was happening, they were joined. According to him, she responded like any other woman. When they finished, they stayed on the couch, petting, and within minutes, both were ready again. He tried to kiss her, but she refused, preferring to nibble his chin. When the second sexual act was over, she began to avoid him. As she stood up, the door opened and one of the alien men stepped in, calling to the woman. Before she left, she smiled at Villas-Boas, pointed to her stomach and then to the sky. "Southward," according to Villas-Boas.

One of the men came back and handed Villas-Boas his clothes. While dressing, he noticed that his cigarette lighter was missing. He thought that it might have been lost during the struggle in the field. Or, it might have disappeared on the ship.

The creature directed him out of the small room and into another room where the crew members were sitting, talking, or rather growling, among themselves. Villas-Boas was left out of the discussion, the aliens ignoring him, so he tried to fix the details in his mind. At last one of the aliens motioned for him to follow. None of the others looked up as he took a quick tour of the ship. The door was open again, with the ladder extending to the ground, but they didn't descend. Instead, they stepped onto a platform that went around the ship. Slowly they walked along it as the alien pointed out various features. Since he didn't speak, Villas-Boas didn't know the purpose of any of the things he was shown. There were machines with purplish lights. He glanced up at the cupola, which now emitted a greenish light and was making a noise like a vacuum cleaner as it slowly spun.

When the tour ended, he was taken back to the ladder, and the alien motioned toward the ground. When Villas-Boas was at the bottom, he stopped and looked back, but the alien hadn't moved. Instead, he pointed to himself, then to the ground, and finally toward the sky. He signaled Villas-Boas to step back as he disappeared inside the craft.

The ladder telescoped back into the craft and the door vanished. When it was closed, there was no sign of a seam or a crack. The lights brightened as those on the cupola began to spin faster and faster until the ship lifted quietly into the night sky.

As the ship disappeared, Villas-Boas walked back to his tractor. He noticed that it was 5:30 in the morning. He had been on the ship for more than four hours.

This is the first reported incident that can be termed a "classic" alien abduction. It was the first time that a serious UFO-investigating organization, APRO, accepted a tale of physical contact with alien beings. Prior to Villas-Boas's account, the only stories of such interaction came from the contactees—a group who claimed detailed knowledge of UFO occupants and trips inside the ships—but that were almost universally rejected. Now, given the spin of the Villas-Boas tale, that he was captured for experimentation, the story of his abduction, for some reason, was more acceptable.

The tale did, of course, foreshadow the story of Barney and Betty Hill, who, in 1961, claimed to have been abducted by aliens for experimentation on board a flying saucer. When that story was finally told publicly in 1966, it wasn't quite as unbelievable as it would have been without the Villas-Boas tale sitting in the files.

In other categories of UFO phenomena, the United States has led the way. The first man killed chasing a flying saucer was Thomas Mantell, although that object was eventually identified as an experimental balloon. During an intercept of a UFO in 1953, pilot Felix Moncla, Jr., and his radar officer, R. R. Wilson, disappeared over Lake Superior. Neither they nor the wreckage of their plane were ever found. It is one of the few tales in the UFO literature in which some kind of alternative explanation has not been found for what happened.

More recently, in Australia, a young pilot, in communication with a ground station, vanished. On October 21, 1978, Frederick Valentich, a twenty-year-old instructor pilot took off from Moorabin, Victoria, Australia, heading toward King Island, Tasmania. He was flying a Cessna 182 designated Delta Sierra Juliet, which was part of its registration number. Radio contact was maintained in routine flight following through the Melbourne Flight Service Unit and controller Steve Robey.

Just after 7:00 P.M., Valentich contacted Melbourne and asked if there was any traffic, that is, other aircraft below him at five thousand feet. The answer was, "No known traffic."

"I am—seems [to] be a large aircraft below five thousand," said Valentich.

"What type of aircraft?"

"I cannot affirm. It [has] four bright . . . it seems to me like landing lights. . . . The aircraft has just passed over me at least a thousand feet above."

Flight following (Robey) said, "Roger, and it is a large aircraft? Confirm."

". . . Unknown due to the speed it's traveling. Is there any Air Force aircraft in the vicinity?"

"No known aircraft in the vicinity."

"It's approaching now from due east towards me." Valentich was silent for several moments and then added, "It seems to me that he's playing some sort of game. He's flying over me two to three times at speeds I could not identify."

"Roger. What is your actual level?"

"My level is four and a half thousand. Four five zero zero."

"Confirm you cannot identify the aircraft," said Robey.

"Affirmative."

"Roger. Stand by."

Valentich then said, "It's not an aircraft. It is . . ." There was a brief silence.

"Can you describe the, er, aircraft?"

"As it's flying past, it's a long shape. [I cannot] identify more than [that it has such speed]. . . . [It is] before me right now, Melbourne."

"And how large would the . . . er . . . object be?"

"It seems like it's stationary. What I'm doing right now is orbiting, and the thing is just orbiting on top of me. Also it's got a green light and sort of metallic. It's shiny . . . the outside. . . . It's just vanished. . . . Would you know what kind of aircraft I've got? Is it [a] military aircraft?"

Robey said, "Confirm that the . . . aircraft just vanished."

"Say again."

"Is the aircraft still with you?"

"Approaching from the southwest. . . . The engine is rough idling. I've got it at twenty-three twenty-four, and this is . . ."

"Roger. What are your intentions?"

"My intentions are, ah, to go to King Island. Ah, Melbourne, that strange aircraft is hovering on top of me again. . . . It is hovering, and it's not an aircraft. . . ."

That was the last message that anyone received from Frederick Valentich. There were, according to a number of sources, an additional seventeen seconds of sound. There was no voice, just metallic scraping. And then there was just dead silence. Repeated attempts by Robey to raise Valentich failed. Valentich, and his aircraft, seemed to have vanished.

When Valentich failed to land at King Island at his estimated arrival time, other light aircraft began to make visual searches but found nothing. Although Valentich's aircraft was equipped with a radio–survival beacon, nothing was heard from it.

A Royal Australian Air Force (RAAF) Orion, a long-range maritime reconnaissance aircraft, conducted a tracking crawl, following the course taken by Valentich's Cessna and continued searching all day Sunday. They continued their search on Monday, and an oil slick was spotted. Samples of the slick, taken by ships dispatched to the area, showed that the material was marine diesel fuel and not aviation gas. Clearly it was not from Valentich's missing Cessna.

After several days, with no positive results, the search operations were quietly suspended. Although they had found debris in the general area of where Valentich and his aircraft would have disappeared, the material was identified as packing cases and plastic bags. It had nothing to do with Valentich or the Cessna.

Australia's Bureau of Air Safety released a report in May 1982, three and a half years after the Valentich event, explaining that they really knew nothing about it. They were unable to determine the location of the "occurrence"; they didn't know the time that it happened; they believed it was fatal but there was no body or wrecked airplane; and they had no opinion as to why the aircraft disappeared. In other words, they knew nothing and added nothing to the case.

There have been, of course, other tales of disappearing aircraft from around the world. The stories illustrate, once again, the worldwide nature of the UFO phenomenon. Another aspect that might illustrate that point better than any other are the photographic cases. Just as there are here, in the United States, literally hundreds of pictures of UFOs, some the classic flying saucer, others cigar-shaped craft, or the nocturnal lights that seem to haunt us, there are similar numbers of UFO pictures from around the world. Some countries are better represented than others, but that just might be a reflection of the size of the country, the population of it, how often and for how long they get outside, how technologically advanced they are, or any number of other variables.

Just as seen elsewhere, the quality of the photographs varies from photographer to photographer. The credibility of the pictures also varies greatly, just as it has in this country. But all this merely underscores the fact that UFO sightings are not limited to one location in the world. They have been, as shown in the following cases, photographed around the world.

"Light Absorbing" UFO:

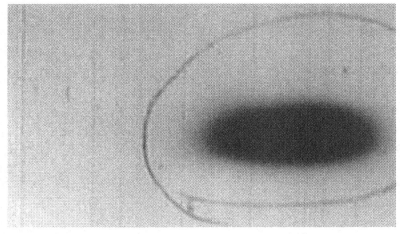

Location: Chitose (U.S.) Air Force Base, Japan

Witnesses: Unidentified U.S. Air Force enlisted man

Craft Type: Darkened disc

Dimensions: Unknown

Primary Color: Black

Sound: None

Exhaust: None

Quality of Photos: Sharp and clear

Number of Photos: 1

Sources: Project Blue Book files

Reliability: 3

Narrative: A U.S. Air Force enlisted man, on temporary duty at the air force base in Chitose, Japan, was taking photographs when he sighted, through the viewfinder, a dark object moving from left to right. Although the object appeared in the photographs when processed, the airman didn't report the sighting until his return to the United States. He believed that the Ground Control Approach radar unit on the base would be able to confirm the sighting, but air force records indicate that they did not.

The air force concluded that the sighting was the result of a processing error or that the object was the result of a negative flaw. An unidentified officer wrote, "In the processing of this film, development in the area of the dark blotch was prevented entirely and in the surrounding area partially, leaving a clear and partially clear area on the film. This area, in printing, allowed the full light of the printer to pass onto the paper and expose it to this degree of black."

Of course, since the object photographed was black, the same thing would have been noticed on the film. In other words, a jet black object would have appeared white on the negative, creating the same effect when making the prints. The air force explanation, in reality, explained nothing.

It should also be noted that this wasn't a case in which the photographer didn't see the object when he took the picture. The witness provided air force investigators with information about the object, including the estimates that it was about one-half to a full mile away and that it was at about one thousand feet. He said that he watched it for about two or three seconds, and that when he moved the camera away from his eye, the object had disappeared.

If the story told by the airman is true, then the air force explanation is false. A processing flaw or negative flaw wouldn't have been visible to the photographer. It would only have appeared once the film had been processed.

If the air force explanation is correct, it would mean that the man who took the picture was lying because there would have been nothing for him to see. The air force apparently rejected the testimony of the witness out of hand, believing him to have fabricated the sighting to explain the negative flaw.

Major Robert Buckmaster, in analyzing the photograph, wrote, "If this were a dark flying object, these things would be true: a. Even if the UFO was jet black as indicated on the print there would be some detail visible in this area of the film, at the indicated exposure. b. There would be a more definite outline to the object. c. If the sky was at all hazy or if the object was a distance away a dark or a light object would have a tendency to blend into the haze or sky."

Buckmaster concluded that the film, during development, touched some part of the tank causing the film to be clear and leaving a black spot on the print. Buckmaster did not believe that an object had been seen in the sky.

The Trindade "Saturn-Shaped" UFO:

JANUARY 16, 1958

Location: Trindade Island, Brazil

Witnesses: Almiro Barauna
Carlos A. Bacellar
About 100 others

Craft Type: Saturn-shaped

Dimensions: 124 feet in diameter, 24 feet thick

Primary Color: Gray

Sound: None

Exhaust: None

Quality of Photos: Sharp and clear

Number of Photos: 5

Sources: Project Blue Book files; *Flying Saucers: The Startling Evidence of the Invasion from Outer Space* by Coral Lorenzen; *The*

World of Flying Saucers by Donald Menzel and Lyle Boyd; *The UFO Enigma* by Donald Menzel and Ernest Taves; *The Emergence of a Phenomenon: UFOs from the Beginning through 1959* by Jerry Clark; *The UFO Evidence* edited by Richard Hall

Relation to Other Sightings: Passo Fundo, Brazil (1976); Saul Janusas (1978)

Reliability: 8

Narrative: Few photographs had as many witnesses as those taken on January 16, 1958, from the decks of the *Almirante Saldanha*, which was part of the Hydrography and Navigation Division of the Brazilian navy. Just after noon, after having been anchored near Trindade Island for several days, the ship was preparing to depart. On deck, someone spotted a strange object in the

A saturn-shaped object flies above the island.

slate-gray sky and pointed it out to the others. It had a flattened, spherical shape with a ring around the center of it. The craft flew over the island, hovered briefly, and then continued on, disappearing to the east.

A photographer, Almiro Barauna, invited on board the ship by the navy to take underwater pictures, was standing on the deck when the object was seen. He would later tell the following to reporters for the Rio de Janeiro newspaper *Jornal do Brasil:*

Suddenly Mr. Amilar Vieire and Captain Viegas called me pointing to a certain spot in the sky and yelling about a bright object which was approaching the island. At this same moment, when I was still trying to see what it was, Lieutenant Homero, the ship's dentist, came from the bow toward us, running, pointing to the sky and yelling about the object he was watching. Then I was finally able to locate the object by the flash it emitted. It was already close to the island. It glittered at times, perhaps reflecting sunlight, perhaps changing its own light—I don't know. It was coming over the sea, moving toward the point called the "Galo Crest." I had lost about thirty seconds looking for the object, but the camera was already in my hands, ready, when I sighted it clearly silhouetted against the clouds. I shot two photos before it disappeared behind Desejado peak. My camera was set at a speed of .125 with an f/8 aperture, and this was the cause of the overexposure error, as I discovered later. The object remained out of sight for a few seconds, behind the peak, reappearing larger in size and flying in the opposite direction. It was lower and closer than before and moving at a higher speed. I shot the third photo. The fourth and fifth shots were lost, partly because of the speed at which the object was moving, and partly because I was being pushed and pulled about in the excitement. It was moving in the direction from which it had come, and it appeared to stop in mid-air for a brief time. At that moment I shot my last photo, the last on the film. After about two seconds the object continued to increase its distance from the ship, gradually diminishing in size and finally disappearing into the horizon.

According to a report appearing in Coral Lorenzen's *Flying Saucers: The Startling Evidence of the Invasion from Outer Space*, Barauna told others that the ship's captain, as well as several other officers, wanted to see if he had gotten any good shots of the object. He developed the film immediately, apparently under the supervision of the various officers including Commander Carlos A. Bacellar.

Within days of arriving back at Rio, Bacellar approached Barauna and asked for copies of the pictures. He took them to the navy, and then returned, asking that Barauna describe to authorities how he had taken the pictures and the events that led up to it.

According to Lorenzen, Barauna was asked to appear at the Navy Ministry where he was questioned by high-ranking officers. Barauna also provided the original negatives to the navy, which were analyzed by the Cruzeiro do Sul Aerophotogrammetric Service. Again, according to Lorenzen, Barauna was told that the negatives were genuine and that no evidence of trick or hoax had been found.

Rio de Janeiro's newspapers carried a number of stories about the photographs and the situation relating to them. *Ultima Hora* reported on February 21 that there were at least one hundred witnesses to the sighting. Barauna told another reporter that "at the end of the meeting, the chief intelligence officer said he was convinced my photos were authentic. He showed me another photo taken by a Navy telegrapher sergeant, also at Trindade."

Commander Paulo Moreira da Silva of the Hydrography and Navigation Service issued a statement on February 22. He said, "The object sighted in the skies of Trindade was not a weather balloon, nor an American guided missile. I cannot yet give my conclusions, for the data are being analyzed in a secret evaluation at the Navy Ministry. I can tell, however, that the object was not a meteorological balloon."

"The object . . . was not a weather balloon."

Dr. Olavo Fontes, APRO's Brazilian representative, visited the Navy Ministry. Everything he learned there convinced him that no hoax was being perpetrated. In fact, Fontes was shown the four pictures taken by Barauna, and a fifth, taken by another photographer days before Barauna photographed the object.

Within days, a Brazilian congressional investigation was launched. According to Fontes' report the details were published in all the Rio de Janeiro newspapers:

The Navy Ministry is requested to answer or explain the following items of the inquiry presented by Representative Sergio Magalhaes on February 27, 1958, and approved by this House: (1) Whether it is true that the crew of the Almirante Saldanha witnessed the sighting of a strange object over the Island of Trindade; (2) Considering that the official statement released from the Navy Ministry's office recognizes that the photos of the strange object were taken "in the presence of members from the crew of the Almirante Saldanha," it is asked whether an investigation was made and whether the reports from the Navy officers and sailors involved were recorded; (3) In the hypothesis of a negative answer the Navy Minister is requested to explain the reasons on which he has based his inclination to attribute no importance to the fact; (4) If it is correct that the photos were developed in the presence of officer from the Almirante Saldanha and that the picture showed the image of the strange object since the first examination; and (5) If the negatives were submitted to a careful examination to detect any photographic trick contrived before the sighting; (6) Why the information was kept secret by Navy authorities for about a month; (7) Whether it is correct that other similar phenomena were observed by Navy officers; (8) Whether it is correct that the commanding officer of the Navy tow ship Tridente witnessed the appearance of the strange object called a "flying saucer."

The appearance of these strange aerial objects known as "flying saucers" has attracted the world's interest and curiosity for more than ten years. For the first time, however, the phenomenon is witnessed by a large number of members from a military organization, and the photos of the object receive the official seal through a statement released to the press by the Navy Minister's office. Yet, as the problem affects the national security, more information is necessary to clarify the facts. There is some controversy in the information divulged through the press, but the Navy apparently has no intention of releasing the complete report to stop the confusion and inform the public. Furthermore, in spite of the Navy Minister's officer declaring (officially) that a large number of people from the Almirante Saldanha crew had sighted the strange object photographed over the Island of Trindade, there was no request for the witnesses' reports or any other measures, as the chief of the Navy high staff admitted when interviewed by the press.

Two months later, the newspapers reported on the official, and secret, Navy answer to the House of Representatives. The reports noted that the investigation hadn't started with the Barauna photographs, but some weeks earlier when UFOs began appearing over the island. The analysis of the Barauna photographs concluded, "Personal reports and photographic evidence of certain value indicate the existence of unidentified aerial objects."

Jerry Clark, in his massive encyclopedia about the UFO phenomenon, provided a slightly different version of the quote: "Existence of personal reports and of photographic evidence, of certain value considering the circumstances involved, permit the admission that there are indications of the existence of unidentified aerial objects." The two versions say basically the same thing.

But this would not be the last word. Donald Menzel soon labeled the case a hoax. Menzel wrote Dick Hall, then with the National Investigations Committee on Aerial Phenomena (NICAP), that he had reached a conclusion about the sightings over Trindade. Menzel wrote, "I have in my possession one well-

Could it be a plane flying through fog?

authenticated case of a saturn-like object, whose nature is known and clearly distinguishable in this particular instance. A plane, flying in a humid but apparently super-cooled atmosphere, became completely enveloped in fog, so about all one could see was a division where the stream lines were flowing up and down respectively over and under the wings. The cabin made a saturn-like spot in the center, and the wings closely resembled the appearance of the Brazilian photographs." The speed and maneuvering, according to Menzel, was the result of an illusion created by the sun shining on the highly reflective fog.

Menzel, apparently unsatisfied with his explanation, came up with another one a few years later. In the book *The World of Flying Saucers,* coauthored with Lyle Boyd, Menzel wrote that the Trindade case was a hoax. Barauna had used a double exposure to fake the photographs and then worked with an accomplice to create a sensation, as well as a believable story. He claimed that when reporters tried to interview the officers and crew of the ship, none of them said they had actually seen the object—this despite the fact Dick Hall had sent Menzel a copy of a March 1958 article from the Brazilian magazine *O Cruzeiro* in which several of the witnesses were named. Lorenzen, in her book, provided additional names of those involved in the case.

Menzel, who always criticized the UFO community for playing fast and loose with the facts, rewrote the Brazilian press release. However, Jerry Clark stated that "when the original and Menzel's reprint are compared, some significant discrepancies become apparent."

The original press release read, "Evidently, this Ministry cannot make any statement about the object sighted over the island of Trindade, for the photographs do not constitute enough evidence for the purpose." Menzel's version reads, "*Clearly,* this Ministry cannot make any statement about *the reality* of the object, for the *photos* do not constitute enough evidence for *such* a purpose [Changes in italics]."

Clark noted, "Whereas the first statement acknowledges an object and a sighting, the second [reprinted by Menzel] implies that their reality is open to questions—hardly the Brazilian Navy's intention."

Menzel, along with coauthor Ernest H. Taves, attacked the Trindade case again in *The UFO Enigma*. They reported that "a hoax notorious in UFO annals was launched in February 1958, when a professional photographer aboard the Brazilian training ship, the *Almirante Saldanha,* took a series of photographs of an alleged UFO flying over the island of Trindade, in the South Atlantic Ocean east of Brazil. Acclaimed by the flying-saucer clan as the most convincing UFO pictures yet available, the photographs were published internationally. The apparent endorsement by the Brazilian Navy and later by the Brazilian government itself quickly collapsed when it was established that the photographer was well known for his trick photography and that no one else, except a friend (and presumed accomplice), had seen the disk flying overhead. It also came to light that the same photographer had, a short time before, attempted to sell a number of fabricated photographs of UFOs."

Menzel and Taves then offer a way for the photographer to have faked the pictures. They theorized that Barauna could have taken pictures of a model against a black background and then reloaded the camera with the same film. Once on the ship, he could photograph the landscape of the island, producing his very convincing fakes. Of course Menzel and Taves offer no evidence that such was the case, only that it could have been done.

But, once again, others simply didn't agree with this assessment. There were, literally, dozens of witnesses, making this one of the few photographic cases in which there were multiple witnesses. The stories told by those witnesses seemed to corroborate one another. And, according to the reports, the film was developed in the presence of the ship's officers.

Menzel had taken the photographs and changed his opinion about them from a rather rare weather-related phenomenon to one of hoax. It was not the first time that he had done this.

Menzel's idea about the Trindade Island pictures might have been the result of U.S. Navy interest in them. They refused to make any public comment, but in 1963, Major Carl R. Hart (not to be confused with Carl Hart, Jr., of Lubbock) did make a statement about the Trindade case in an Office of Naval Intelligence report. He wrote, "This gentleman [Barauna] has a long history of photographic trick shots . . . he prepared a purposely humorous article, published in a magazine, entitled 'A Flying Saucer Hunted Me at Home,' using trick photography." But the article had been written as a debunking piece to show how other UFO photographs, especially those taken in Brazil in 1957 could have been created.

Dr. Mark Rodegheir of the Center for UFO Studies said that Hynek had gone to Brazil in the early 1970s, talked with Barauna and some of the other witnesses, but no photometric examination of the photographs was attempted. Here was an opportunity to advance the case for the existence of UFOs, or at the very least, provide a form of physical evidence proving that something was happening and something was being observed. But the climate of the times, the distances involved, and our own attitudes toward both UFOs and South America have colored our thinking.

A Cuban Cigar:

Location: Guantanamo Bay, Cuba

Witnesses: U.S. Army private

Craft Type: Cigar-shaped

Dimensions: Unknown

Primary Color: White

Sound: None

Exhaust: None

Quality of Photos: Sharp and clear

Number of Photos: 1

Sources: Project Blue Book files; *APRO Bulletin*

Reliability: 4

Narrative: The air force file on the Guantanamo Bay case provides little in the way of information. Colonel Philip G. Evans reported, "Analysis of the photograph indicates that the photo was probably exposed inside of a building and through a glass window. The object in question is probably a reflection on the window from a light hanging from the ceiling. The inside of the shade creates an elliptical shape and the strong white area in the top central portion is the light bulb which also gives off a halation creating the 'bump' on the top of this reflection."

There is no other information in the file. A number of air force forms, which were usually sent to witnesses, are not there. The name of the witness has been deleted, so additional information is impossible to obtain.

It should be noted that APRO's Coral Lorenzen received a copy of the picture at some point with almost no information about it. She published it in the *APRO Bulletin,* asking for more background on the photo. She speculated that it had been taken in the Panama Canal.

An Italian Hoax:

SEPTEMBER 26, 1960

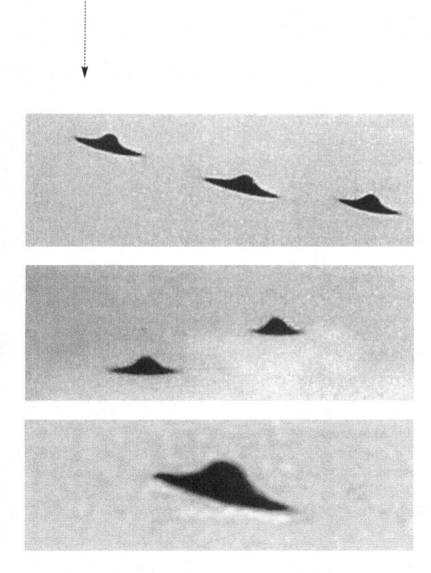

Location: Italy

Witnesses: Unidentified teenager

Craft Type: Domed disc

Dimensions: Forty-five feet in diameter

Primary Color: Black

Sound: None

Exhaust: None

Quality of Photos: Sharp and clear

Number of Photos: 3

Sources: Project Blue Book Files

Reliability: 3

Narrative: The photographs, from an unidentified Italian photographer were submitted with a letter, not to the air force but to NASA. According to the letter, translated from its original Italian, the witness noted, "I am taking the liberty of writing you to call your attention to the enclosed photographs taken by a friend of mine on 9/26/60 at around 2 p.m. The shape of the object was round, about 15 meters in diameter. Photographed with a shutter of 6.5, time 1/250, distance infinite."

Air force analysis suggested that the case was a probable hoax because there was "too little information to allow valid conclusion. Photo analysts state: objects either very small and close to camera or moving rapidly due to their being out of focus. Due to the objects being much darker than other dark areas, negative or original print was possibly retouched. In one print, objects' size increased, indicating they were closer to camera than in other print. However, other objects in print also increased in size, indicating that the camera was moved toward the background. Also in this print, background appears overexposed, but objects are just as dark. Probably result of someone attempting a hoax."

We suppose we should point out the obvious contradiction in the air force analysis. They note there is not enough information to draw a "valid conclusion" and then do exactly that. Although there seems to be little doubt that the photographs are the result of a hoax, the analysis by the air force suggests something about the validity of their investigations.

One from Sweden:

MAY 6, 1971

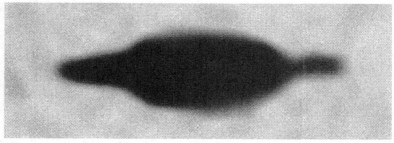

Location: Sweden

Witnesses: Lars Thorn

Craft Type: Disc

Dimensions: Thirty-five feet in diameter

Primary Color: Black

Sound: Whizzing at ten- to fifteen-second intervals

Exhaust: None

Quality of Photos: Sharp and clear

Number of Photos: 2

Type of Camera: Minolta

Sources: *The APRO Bulletin,* January/February 1972

Reliability: 4

Narrative: APRO reported that Lars Thorn snapped two pictures of a UFO as it hovered near a bunker in the middle of Sweden. Thorn said that it rocked from side to side, allowing him to get a very good look at it. He said that on the upper side there was a slight dome, and just below the dome was an area of red and gray and below that area was a band that looked like a green ribbon.

Thorn said that he was riding his motorbike when, through the trees, he spotted the object. It traveled to a position near the bunker, stopped to hover, and then sped off in the direction from which it had come. While it was hovering, Thorn took the photographs of it.

According to APRO, the photographic negatives were examined by experts who performed a number of tests on them. They looked for any evidence that there was a string or thread holding up a small model or that it was some sort of composite print but could find no evidence of either fabrication.

Other experts claimed that the picture was of a model held aloft by a hot air balloon. One expert suggested that under magnification, he could see a line running from the ground to the bottom of the object, suggesting, to him, a tether. No one else was able to see any evidence of that or of a hoax.

This is a case in which the object was either a UFO or a hoax. There was no middle ground in which some kind of ambiguous light source was photographed. To date, no one has proved the sighting a hoax, and Thorn has said that he faked nothing.

More from Brazil:

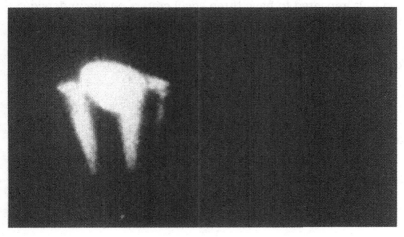

Location: Rio de Janeiro, Brazil

Witnesses: Two unidentified female students
Nelson Calmon Schubsky
Sheyla Fernandes Cardoso

Craft Type: Sphere

Primary Color: Rose-colored with white, yellow, and red lights

Craft Description: It was seen as bright lights with an object behind it. It appeared to have three small craters on the underside. Three lights shined from the underside toward the ground.

Sound: None

Exhaust: None

Quality of Photos: Sharp and clear

Number of Photos: 2

Type of Camera: Leica Model IIIf

Sources: *The APRO Bulletin,* January/February 1972

Reliability: 6

Narrative: According to the *APRO Bulletin,* two young female students were among the first to see the object and called to others to come look at it. Hearing the shouts from outside, Nelson Schubsky and his fiancée, Sheyla Fernandes Cardoso, rushed into the street to find a large number of people observing the object. Schubsky was carrying his camera with him.

He believed he had no time to adjust the camera, other than to open the diaphragm to its maximum. He took two photographs of the object as it maneuvered near a factory. As it moved, it seemed to pulsate, the color deepening and then fading.

According to the witnesses, the object finally disappeared behind a chimney on the factory and did not reappear. Those on the other side of the factory said they saw a red glow but saw no object.

There is one point that needs to be made here, and that is a seeming contradiction. Shubsky said that he had no time to adjust the camera properly before he began to take the pictures. He shot two photos, and eventually the object disappeared. It might be unfair to discuss this from the point of view of someone not on the scene, but five minutes is quite a long time. Schubsky could easily have taken a couple of shots and then attempted to adjust the camera to refine the photographs. However, there is nothing to suggest that he took more than the two pictures published by APRO.

A Greek UFO:

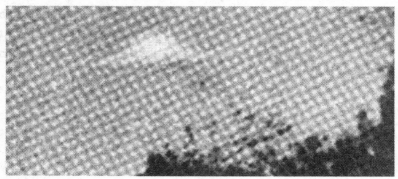

Location: Athens, Greece

Witnesses: Unidentified

Craft Type: Domed disc

Dimensions: Forty feet in diameter

Primary Color: Gray

Sound: None

Exhaust: None

Number of Photos: 1

Type of Camera: Russian-made Lubitel 6x6

Sources: *The APRO Bulletin*, March/April 1974

Reliability: 6

Narrative: As has happened on a number of occasions, the object that appears in the photograph from Greece was not seen when the picture was taken. According to APRO field investigator Dimitrios Zambicos, the photographer was taking pictures of a street, with the camera mounted on a tripod. He took the picture and then found the object when the film was developed.

According to the *APRO Bulletin,* the camera was pointed to the northwest and the sky was overcast. About ten minutes after the picture was taken, snow began to fall.

A Canadian Craft:

MARCH 18, 1975

Location: Hamilton, Ontario, Canada

Witnesses: Pat McCarthy

Craft Type: Domed disc

Dimensions: Twice the size of a DC-8

Primary Color: Black

Sound: None

Exhaust: None

Quality of Photos: Sharp and clear, with no foreground detail

Number of Photos: 3

Type of Camera: Praktica Nova 1 with a Hanimar 135 mm lens

Sources: *APRO Bulletin,* October 1975

Relation to Other Sightings: Helio Aguiar Photographs (April 1959)

Reliability: 4

Narrative: Pat McCarthy had gone out in the hopes of photographing hawks near his home. He failed to get those photographs, and as he was returning home, he caught a movement out of the corner of his eye. When he turned, he saw an object, which to him, resembled a Frisbee.

McCarthy said that the object was moving too fast to be a bird and he decided to photograph it. He said that it was moving too fast for him to keep it in the viewfinder, but he finally snapped a picture. He tried a second shot, thought that he missed, and took a third. He lost sight of the object momentarily, and then took a fourth photo. He didn't take a fifth because the object was moving out of his view at great speed. It disappeared into the distance.

McCarthy wasn't certain that he had caught any images on the film until it was developed. Worried that the pictures would be called a hoax, he took the film to the local newspaper to use their darkroom, even though he had his own. When he developed the film he discovered that he had, in fact, missed the second shot. All that showed in it was a tree.

McCarthy, who was described as an amateur astronomer, and said that he was familiar with various types of aircraft, told investigators that the object seemed to be twice the size of a DC-8 and that he believed, based on the cloud formations, the object was between fifteen and twenty thousand feet.

APRO researchers were excited about the case because it seemed to match a series of photographs taken in 1959 at the Piata Beach in Bahia, Brazil. However, although it is clear that the Bahia photographs match McCarthy's in some respects—generally, the domed discs and the positions they were in when the respective pictures were taken, the objects themselves don't match fully.

Back to Brazil:

Location: Passo Fundo, Brazil

Witnesses: Joshua da Silva
Gesareo Goncalvas

Craft Type: Saturn-shaped

Dimensions: Twenty-five feet in diameter

Primary Color: Gray

Sound: None

Exhaust: None

Quality of Photos: Sharp and clear

Number of Photos: 2

Type of Camera: Kodak Rio

Sources: *APRO Bulletin*, November 1978

Relation to Other Sightings: Trindade Island (1958); Saul Janusas (1978)

Reliability: 4

Narrative: Gesareo Goncalvas was driving when Joshua da Silva spotted a metallic object in front of them. Goncalves drove forward slowly until they were just over a hundred feet from the object. They believed it to be made of metal and thought that it was about twenty-five feet in diameter.

When the car stopped, da Silva took one photograph, rolled the film forward and took another. At that point, the object began moving to the northeast, increasing its speed until it disappeared. It vanished in a matter of seconds.

According to APRO, the negatives were examined by technicians at *El Globo,* one of Rio de Janeiro's newspapers. They said that they found no evidence of a hoax.

Still More from Brazil:

Juпе 20 oʀ 21, 1978

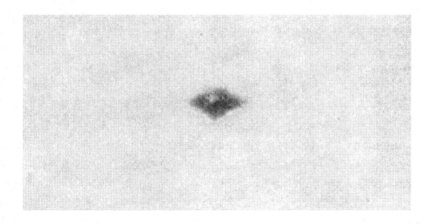

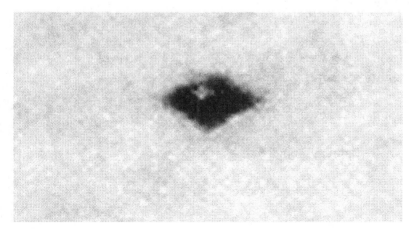

Location: Rio de Janeiro, Brazil

Witnesses: Saul Janusas

Craft Type: Saturn-shaped

Dimensions: Thirty feet in diameter

Primary Color: Gray

Sound: None

Exhaust: None

Quality of Photos: Sharp and clear

Number of Photos: 2

Type of Camera: Kodak Rio

Sources: *APRO Bulletin,* November 1978

Relation to Other Sightings: Trindade Island (1958), Passo Fundo (1976)

Reliability: 4

Narrative: After arriving at Rio de Janeiro International Airport, Saul Janusas, an electrical engineer from New York, had boarded a bus for a ride to Copacabana. Not far from the airport, he saw an object through the bus window. He grabbed his camera, took a photograph, and then wound the film to take a second picture.

Before he could take a third picture, the bus turned and Janusas lost sight of the object. He didn't tell anyone on the bus about the object because they all spoke Portuguese, and he didn't learn if there were any other witnesses.

The Belgium Triangle Begins:

November 29, 1989

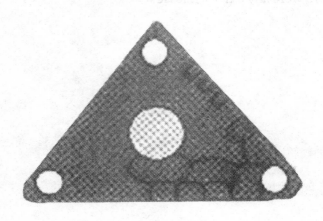

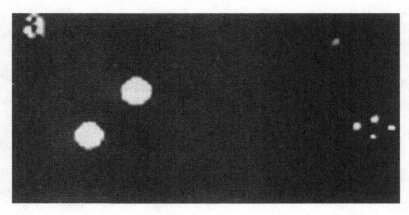

Location: Belgium

Witnesses: Francesco Valenzano
Andre Amond
More than a hundred others

Craft Type: Triangular

Primary Color: Gray

Sound: Periodically reported roar or drone

Exhaust: None

Sources: *Project Moon Dust* by Kevin Randle; "The Belgian Wave" by Patrick Vidal and Michel Rozencwajg (in *International UFO Reporter,* July/ August, 1991); "They Came from Outer Space and Belgium was Their Target" (in *The Bulletin,* January 11, 1996); *UFO: The Complete Sightings* by Peter Brookesmith; *Alien Contact* by Timothy Good

Reliability: 4

Narrative: All the accounts of the Belgium wave of UFO reports begin by mentioning the first sightings of a triangular-shaped object on November 29, 1989. More than 120 people including thirteen police officers were among the witnesses present at the Eupen, Belgium, sighting. The object, with shining lights, was seen at close range. Although the sighting had multiple witnesses, what they saw was little more than nocturnal lights. It did, however, mark the beginning of a series of sightings that would produce some interesting and impressive evidence.

Just two days after the Eupen sighting, Francesco Valenzano was driving through Ans in the Liege province of Belgium, when his daughter shouted that he should look up. In the sky above them was a large, triangular-shaped object that drifted along slowly, just above the buildings of the town square. Finally it flew off toward another village.

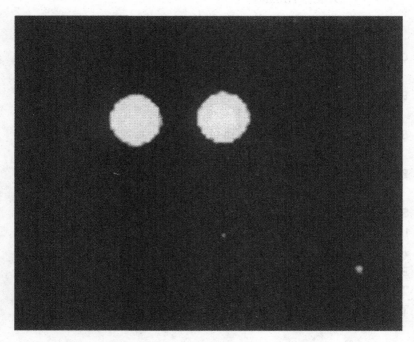

Many of the sightings were of nocturnal lights.

There would be more sightings and the list of witnesses would continue to grow. Valenzano was a Belgian air force meteorological specialist. On December 11, a Belgian army lieutenant colonel, Andre Amond, while driving with his wife, spotted a slow-moving strangely lighted object moving at low altitude. Amond stopped the car so that he could see the object better. It flashed a beam of light at him and began to approach. His wife began to shout, wanting to get away from the object.

Amond said that the object then departed rapidly. He later told investigators that he had been impressed by the slowness with which the object moved originally, and then its speed as it flashed away. The object made no noise as it maneuvered.

The day after Amond's sighting, and in response to the growing wave of UFO reports, the Eupen police and the Belgian Society for the Study of Spatial Phenomena (SOBENS) held a press conference in Brussels. Also at the press conference were Belgian air force officers who would be investigating the sightings and Guy Coeme, the Belgium defense minister.

No explanations for sightings were offered. The purpose of the conference was to inform the media about the sightings and let them know what was to be done. It was the first time that anyone from the Belgium Ministry of Defense had spoken out on the topic of flying saucers.

Not long after the press conference, one of the daily newspapers in the Brussels area printed a story that suggested the UFOs were actually American-flown F-117As. The article reported that there were secret test flights being conducted and that this particular explanation for the UFOs had come from Washington. Although the F-117A bore a slight resemblance to the triangular UFO, it could not drift along slowly, and the pattern of lights reported didn't match that on the aircraft.

Auguste Meessen, a professor of physics at the Catholic University at Louvain, investigated this explanation, learning that it hadn't come from Washington as the reporter had claimed, but from Finland. A writer for *Het Laatste Nieuws* said that he had just read an article about the F-117A and had wanted to inform his readers about the strange aircraft. To make the article interesting, he had suggested that the recent UFO sightings in Belgium might have been caused by the Stealth Fighter. He had no inside knowledge, just his wild speculation. As a result, other newspapers had grabbed the explanation and ran with it rather than checking it first.

Lieutenant Colonel Wilfried De Brouwer, the Chief of Operations of the Belgium Air Force, told Meessen that they had sought help from the American embassy in an attempt to find solutions for the sightings. Had the F-117A been the culprit in the sightings, De Brouwer would have been told about the flights prior to their being made. The U.S. Air Force does not routinely invade the airspace of a friendly foreign nation without alerting that nation to its presence.

But even the negative press in Belgium didn't discourage more sightings and more reports. On December 12, 1989, a witness, who wished to remain unidentified, was awakened by a throbbing sound. He believed that a circulation pump was about to fail, so he picked up a flashlight and headed toward the boiler. When he shut down the pump, he could still hear the noise. It had to be coming from outside.

As soon as he stepped out, his attention was drawn to an oval-shaped object between two fir trees. In the bright light of the moon, he could see the object clearly. There were small lights around the perimeter that changed color from blue to red and back to blue. The object was metallic-looking. At the front was a window or porthole of some kind.

After a few minutes, the object rose slightly and the sound coming from it changed. It began to drift toward a meadow, shining searchlight-like beams under it. The object then disappeared behind a house but a moment later a well-defined bright light shot into the sky. The witness was too frightened to investigate. Instead he returned to his own house.

Unlike so many others who saw the strange objects, this man reported his sighting to the police. A search of the fields conducted by the police, and the Belgium Army, turned up a giant circular area. The grass had been cut as if by a lawn mower. There were also traces of some sort of yellow material on the grass.

There were others who later claimed that they had heard the noise from the craft but didn't go out to investigate. A reporter living in the area said that he had been awakened by a bright light outside. He thought it was one of his own lights and rolled over to go back to sleep.

The most spectacular of the sightings, and the one that received the widest press dissemination, occurred on the evening of March 30 and the early morning of March 31, 1990. It had been suggested that when there were reliable sightings of the triangular-shaped object, the Belgian air force would respond with American-made F-16 fighters. When several police officials and a host of other civilian witnesses reported seeing the object, the fighters were dispatched.

According to one account, though the fighter pilots made no visual contact with the UFO, the onboard radars did "lock on" to a target. It moved slowly, only about twenty-five miles an hour but then would accelerate at fantastic speeds. It was reported to have dropped from an altitude of 7,500 feet to 750 feet in about one second.

The fighters locked on to the target three times that night, but each time the UFO evaded them. Eventually both the UFO and the fighters left the area. SOBENS representatives are still unsure of what happened that night but did express praise for the Belgium Air Force for its honesty in reporting the incident.

That wasn't, of course, the end of the sightings in Belgium. Although less than productive, from an evidence viewpoint, reports of those sightings did call attention to the events involving the UFOs. Other sightings were reported during the next several weeks.

A month later, for example, there was a close approach of an object near Stockay, Belgium. On May 4, 1990, a respected archaeologist walked outside to close his greenhouse door and heard the dogs in the neighborhood howling. As he returned to the house, he saw a huge object with a clearly defined outline in a field about five hundred feet from him.

He tried to alert his neighbors, but they were not home. Instead he told his wife and together they saw, at the front of the house, a cone-shaped object with a top that was mushroomlike in appearance.

After watching it for a number of minutes, his wife returned to the house. The archaeologist went with her and then decided to take another look. He walked along his field and saw the object was only 100 to 150 feet away.

The object made no noise. The bottom was bright and opaque, the center was white, and the edges were yellowish. The craft seemed to be twenty-five to thirty feet in diameter at the base and was about fifteen to eighteen feet high.

The mushroom tip at the top of the object detached and began to climb. As it did, it turned a bright orange. After a few moments, the mushroom descended, reattached itself at the top of the craft, and the colors returned to their original shades.

Unable to find any other people to witness the event, the archaeologist returned to his house where his wife waited. They watched the UFO for a few more minutes, then went back inside.

Their son went to the spot of the encounter the next day. He found and videotaped four circular ground traces in a rectangular shape about twenty-five by thirty feet. The grass in each of the circular marks was twisted and depressed. A fine yellowish powder was found on the blades of the grass.

In October 1990, with the Stealth Fighter explanation dismissed by uncooperative Belgian and American officials, a new explanation was trotted out by a French magazine. Now, according to the magazine, it wasn't a stealth aircraft, but some other experimental aircraft, flown by American pilots without notifying the Belgian air force that such flights were taking place. Of course, there is absolutely no evidence that any American aircraft, stealth, experimental, or otherwise, were involved in the Belgian sightings. The

French were just not willing to let go of what, to them, was a good explanation for the sightings.

In fact, in October 1990 there were more and better sightings taking place. On Sunday, October 21, two residents of Bastogne were driving home late when two lights appeared, descending toward them. Echoing the words shouted by another Belgian witness, one of them screamed, "They're coming for us!"

They lost sight of the object for a few moments, but it reappeared, now behind a hedge bordering the right side of the road. Now they thought it might be a reflection on the car window, but when the window was cranked down, the light was still there, now less than fifty feet from the car.

As they drove along, the object seemed to pace them, staying with the car. They slowed and stopped, as did the object. When they accelerated, the light did the same. When they reached the end of the roadside hedge, they stopped. Now they could see a dark mass, more than forty-five feet in diameter. It climbed rapidly and silently into the sky. On the bottom was a ring of seven or eight lights.

Less than forty-eight hours later, more UFOs were seen. At 5:30 in the morning on October 23, a young woman identified only as Regine was awakened by her alarm clock. Through the window of her residence on the outskirts of Athus, near Luxembourg, she saw two bright lights hovering over a hill about a quarter of a mile away. Between the two bright lights was a smaller, dimmer blue light.

Ten minutes later, the lights rose into the sky and began drifting toward her silently. The lights passed over her house, and she ran into the dining room where she could see the object again. She noticed that there was also a small red light on the lower part of the craft.

The object kept moving in the direction of Athus. It then veered to the left, in the direction of Luxembourg and disappeared. There were apparently no other witnesses to the event.

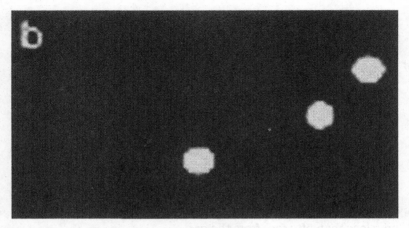

The lights continued to plague Belgium.

Later that same day, that is, October 23, four teenagers saw an object that came from behind a hill. It was about a quarter mile away and had a number of extremely bright lights on the lower part of it that were directed toward the ground. Centered among these lights was a smaller, dimmer light. They believed that the object was taking off, because of its low altitude when they first spotted it.

This craft was shaped like a pyramid, with the apex pointing to the front. At the base were two more red lights. The object was in sight for about thirty seconds and disappeared behind another hill.

Another sighting was made of a craft quite close to the ground. On November 22, 1990, at the village of Fluerus, a young woman was lying in bed when an intense light penetrated the room. She got out of bed, wiped the frost from the glass, and saw that the bright light was coming from behind a neighbor's wall. Although the light wasn't more than a hundred feet away, it was just behind the wall and in an abandoned field. As the light dimmed, she noticed blue flashes from that area. When a train rumbled through, the last of the lights dimmed and vanished.

The sightings continued. On December 9, more than a year after

244

the first of the reports, another couple traveling by car saw an object. At first it seemed to be a glowing triangle. Then, watching through one of the car's windows, they saw a huge, circular plate over the tops of a group of trees. Around the edge were bright white lights. There were also four spokelike lines that were a glowing brass color.

The UFO was about 150 feet in the air. Although they wanted to stop the car, they were unable to do so because of traffic. They lost sight of the object quickly.

The end of 1990 brought no respite. On January 6, 1991, two separate groups of witnesses saw a low-flying circular plate. The first group also reported they had seen a cupola on the top. There were many lights on the object.

Minutes later, another group reported they saw three lights to the left of the road. They, at first, believed it to be the lights from a soccer field. They noticed, however, that the lights were on something hovering over a quarry.

The UFO was, according to one witness, about 250 feet long and 40 to 50 feet high. The underside bulged outward and was dark gray. Fifteen portholes on the side were lighted. The rear was flat and seemed to have some fins.

The press lost interest in the sightings as they continued. The stories were all becoming the same. Witnesses were seeing brightly lighted, mostly triangular-shaped objects floating silently above the ground. There were no real examples of physical evidence left behind, and the few photographs taken provided no proof of the visitations.

The sightings, however, continued, but apparently at a slower pace. SOBENS reported that the sightings were not just of lights in the night sky. These sightings were of objects quite close to the ground with the witnesses, in many cases, not far from the object. The Society was receiving good, detailed descriptions. The only difference was that the media were no longer interested in the reports.

But the blame for the decrease in attention doesn't rest wholly on the media. Those doing UFO research are guilty of sloppy work.

Many reporters have been impressed with the preliminaries of a story only to learn, through investigation, the case isn't as solid as originally portrayed.

This, however, does not excuse sloppy reporting by the media. To suggest, as one Belgium newspaper did, that Washington had "confirmed" the reports of stealth aircraft in the vicinity, is reckless. Clearly there was no such confirmation The reporter should have checked the information, rather than accepting it as true because it was what he wanted to believe.

The one thing that came from the Belgium sightings was governmental cooperation. Rather than bury the details in a classified study that would be released in parts over the next decades, the Belgium Air Force was quite candid. They assisted researchers in trying to find explanations for the series of sightings.

But Belgium wasn't the only location reporting the bright-lighted objects close to the ground. On October 14, 1990, while the Belgium wave was in full swing, a witness in Switzerland saw a similar object. According to Auguste Meessen, writing in the *International UFO Reporter,* Mrs. Wengere and her husband were driving toward Zurich when she spotted two bright white lights.

At first she believed them to be the lights on electrical transmission towers on the mountains. But there was still some sunlight left, so she then thought that the lights were not on the mountains but between the car and the mountains. She told her husband that she was looking at a UFO and asked that he stop their car.

Unfortunately the road conditions wouldn't allow him to stop. They continued their observations, commenting that the lights must be huge because they washed out the stars. An airplane appeared, but the pilots seemed unaware of the UFO, and the object didn't react.

The Wengeres drove into a village and lost sight of the lights. However, when they were back in the country, the lights reappeared, a little higher in the sky. Now a third light had appeared. At first it was motionless and then began drifting up toward the other two lights.

Now the other two lights began to move, as if attached to and circling a single point. The lone light orbited in the same fashion, around some central point, maintaining its distance from the other two lights.

The Wengeres slowed so that they could see the lights better, but there was now traffic behind them. They entered another built-up area and lost sight of the lights, this time for good.

The real point here, however, is that we had a wave of sightings in Belgium in the early 1990s. Hundreds saw the craft and reported them. Belgian air force fighters attempted one intercept. The evidence for the sightings is the same as it has always been: witness testimony. Can the judgments of these witnesses be trusted, or is there something that creates a mob psychology so that normally rational people believe they are seeing flying saucers in the night sky?

The sightings, according to SOBENS have slowed and virtually disappeared. There are a few still made, but not with the regularity of the early part of the decade. What is left is a core of interesting reports that suggest something invaded the airspace over Belgium, and other countries, but there is no evidence that the craft were real or extraterrestrial.

From the Former Soviet Union:

FEBRUARY 4, 1990

Location: Ukraine

Witnesses: Dima Girenko

Craft Type: Domed disc

Sound: None

Exhaust: None

Relation to Other Sightings: Yorba Linda, California, 1967

Reliability: 2

Narrative: The Ukraine sighting, allegedly investigated by the Moscow Aviation Institute, which claimed that it was no hoax, is a UFO report that could have easily been faked. No other information is available.

Part V:

THE POST-BLUE BOOK WORLD, FROM 1969 TO THE PRESENT

With the demise of Project Blue Book in 1969, there was no longer an official entity conducting the investigation of UFOs. Those who called air force bases with sightings were told to contact a local university, or if the caller felt threatened, to alert the local police or sheriff. Filling the gap in the late 1960s and into the 1970s were private organizations such as the Aerial Phenomena Research Organization (APRO) in Tucson, Arizona, and the National Investigations Committee on Aerial Phenomena (NICAP), which was headquartered in Washington, D.C.

APRO was founded in January 1952, by Leslie James "Jim" Lorenzen and Coral Lorenzen. It was designed, at first, as more of a club than a research organization. Coral Lorenzen herself said, "It wasn't really a research group. . . . We adopted that name because the Air Force was putting out those stupid explanations for incidents that were really unexplainable, and I thought there should be an organization that recorded the sightings for later, more responsible scrutiny."

Beginning in June 1952, the group started publishing *The A.P.R.O. Bulletin*. It began with editorial commentary about the state of UFO research and evolved into a newsletter that contained the latest in UFO information. They continued publishing the *Bulletin* for about thirty years, though by the end of the run, the *Bulletin* was appearing sporadically and was filled with typographical errors and poor writing.

During the 1950s—along with the Civilian Saucer Intelligence of New York and the Civilian Research, Interplanetary Flying Objects, two other private organizations—APRO and the Lorenzens believed that a campaign for Congressional hearings was a waste of time. APRO felt that their organization, as well as the other groups, should concentrate on the investigation and documentation of UFO sightings and reports. Let someone else worry about convincing Congress to hold hearings.

APRO also collected sightings that involved the occupants of the crafts and was among the first organizations to report tales of alien abduction. They did not endorse the claims of the contactees who told stories of beautiful alien creatures and visits to distant planets sometimes outside our solar system. They did, however, endorse the tales of those claiming to have been dragged into a spacecraft involuntarily for examination at the hands of the alien scientists.

APRO continued to operate into the 1980s, but the health of both the Lorenzens began a decline. Jim Lorenzen died of cancer on August 28, 1986, and Coral died less than two years later, on April 12, 1988. With its driving force gone, the APRO board decided to dissolve the organization. The files, so painstakingly collected by the Lorenzens during APRO's operation, fell into the hands of a contactee-oriented couple who planned to exploit them for profit-making ventures. Currently the files are disintegrating with no sign that anyone will gain access to them.

The National Investigations Committee on Aerial Phenomena grew out of an idea by the publisher of a contactee-oriented newsletter and a physicist who believed that space flight could be developed by studying UFOs. Within months of its beginning in 1956, NICAP was on the edge of collapse because of money problems and internal disagreements. In January 1957, the old guard was voted out and a new board of directors was voted in.

NICAP believed that the solution to the mystery of the UFOs

would be found through congressional investigation and the penetration of air force secrecy. They built a membership that numbered about 14,000 at the organization's height, but the report authored by Edward U. Condon and his staff at the conclusion of their "scientific" investigation conducted at the University of Colorado—which claimed that UFOs deserved no further investigation—in combination with the end of the official air force investigation, caused a rapid decline in NICAP membership.

NICAP did collect and investigate sighting reports but steered clear, for the most part, of the contactee tales and the accounts in which the occupants of alien craft were seen. Many of the sighting reports were well investigated by competent researchers and proved, at least to the NICAP membership, that something strange and extraterrestrial was being detected.

Although the organization limped along through the 1970s, the membership continued to decline and their newsletter, eventually known as the *UFO Investigator* ceased publication in 1980. The organization was finally disbanded and the files, eventually, were turned over to the J. Allen Hynek Center for UFO Studies (CUFOS).

CUFOS grew out of J. Allen Hynek's "invisible college" of scientists who were quietly interested in UFOs, Hynek's dissatisfaction with the air force's official investigation, and Sherman Larson's belief in UFOs. Larson tried to interest Hynek in creating a group that would draw scientists in and would publish a more scientifically oriented periodical that would provide information to others. Although Larson had already established the Public Education Group, Hynek didn't like the name. He proposed the Center for UFO Studies.

CUFOS was formed shortly after a October 1973 conversation between Larson and Hynek. At first CUFOS headquarters were in Hynek's home, but later they moved into an office, eventually landing on Peterson Avenue in Chicago. In addition, the Center employed a full-time investigator and published various special reports dealing with specific aspects of the UFO field.

Although there were rough times in the 1980s because of a decline in both UFO reports and interest, CUFOS survived. Hynek, due to deteriorating health, moved to the Southwest. Hynek died on April 27, 1986, and Mark Rodeghier took over as scientific director of the Center. To honor Hynek, the name was changed to the J. Allen Hynek Center for UFO Studies.

The Center's major rival, or to be more precise, its major counterpart in the UFO field today, is the Mutual UFO Network (MUFON), started by Walt Andrus. It began as the Midwest UFO Network in Illinois before eventually moving its headquarters to Seguin, Texas. It is the largest of the UFO groups in the country and currently claims a membership of about 3,500. It publishes the monthly *MUFON UFO Journal* and holds an annual convention that brings together the leaders in the field to discuss their various theories and ideas with those who are interested in the topic of UFOs.

Both the Center and MUFON investigate UFO sighting reports and have done so since the beginning of their existence. Both, as noted, publish a periodical. CUFOS also publishes, annually, their *Journal of UFO Studies,* a referred scientific publication. And both the Center and MUFON provide spokespeople to the national media in an attempt to provide rational examination of the UFO field.

The existence of these organizations means that while there is no official government agency to gather UFO related material, there are a number of civilian agencies to do that. Photographic examination is conducted by those with expertise in physics, optics, and various photographic processes. As happened when Blue Book was in operation, dozens of photographs showing a variety of craft have been submitted to the civilian organizations. Some are very interesting.

Noise from a UFO:

JUNE 28, 1967

Location: New Castle, Pennsylvania

Witnesses: Gabriel Kozora

Craft Type: Disc

Primary Color: Gray

Sound: The "sound of a hundred jets"

Exhaust: None

Quality of Photos: Fuzzy

Number of Photos: 2

Type of Camera: Polaroid

Sources: *APRO Bulletin,* July/August 1967; *Flying Saucers, UFO Reports* published by Dell (1967)

Relation to Other Sightings: Muskogee, Oklahoma (1953); Gulf Breeze, Florida (1987)

Reliability: 5

Narrative: While taking pictures of his son, Gabriel Kozora saw an object flying toward him from the south. When it stopped to hover briefly, he took one picture. As it headed toward the northwest, he took a second picture. It eventually disappeared in the north.

When it was learned that Kozora had taken the photographs, he was asked to appear on television to describe what he had seen. Kozora, at first, refused. Later, however, he did appear on a Pittsburgh television station.

The Amana Streak:

Location: Near Middle Amana, Iowa

Witnesses: Mark Leonard

Craft Type: Nocturnal light

Dimensions: Craft seen only as a light

Primary Color: Orange

Sound: None

Exhaust: None

Quality of Photos: Sharp and clear

Number of Photos: 3

Sources: UFO Archives

Reliability: 5

Narrative: High-school student Mark Leonard was attempting to take time exposures of the moon on the evening of November 22, 1975, when his attention was drawn to a bright light overhead. He thought it would be a good reference point for a shot across the pond and centered the light in the viewfinder and snapped the shutter. When he looked through the viewfinder again, he saw that the object had moved. He centered the light in the viewfinder again and took the second picture. After that exposure, he saw that the object had moved so far, he had to move the camera so that he could take the third, and last, of the photographs. The object finally moved behind some trees and was no longer in view.

Leonard said that the light seemed to flicker as it moved, not unlike the way a railroad engine's front light sweeps from side to side. He believed that it was accelerating to the north as it disappeared.

Leonard was quick to show investigators all the negatives he had taken that night. There was no evidence that he had been experimenting with trick photography. The film seemed to bear out the tale he told.

Plotting the flight path on a map revealed that it seemed to be flying too slowly to be meteors or even aircraft. At ten miles' distance, the object would have been moving at only sixty miles an hour, and it is unlikely that the object was that far away. Operations at the nearby Cedar Rapids airport had been suspended for the night by the time the object was photographed. That certainly ruled out commercial aircraft.

Although an explanation for the sighting has been found, it is one that is somewhat speculative. The speeds were plotted, assuming that the object was flying perpendicular to the camera. If, however, it was flying at an angle away from the camera, the speed computations would be flawed. Leonard said that he heard no sound of an engine, but with the wind blowing away from him, he might not have heard it.

Standard computer techniques revealed an object at the head of the streak of light.

The object of the sighting was most likely a small private aircraft heading either to the Cedar Rapids airport or to one of the other small airfields in the area. The weather was fair, though cold. Given that factor, the object could have been a private aircraft operating under visual flight rules with no flight-plan files so that there was no way to discover, when the investigation began, if such a flight had taken place.

A Minnesota Sighting:

NOVEMBER 26, 1975

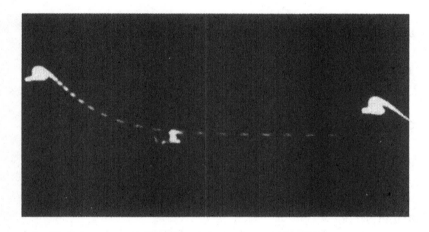

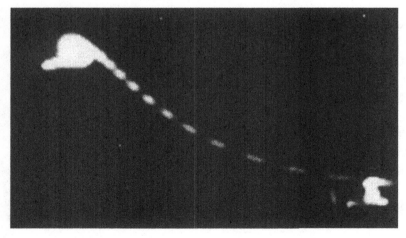

Location: Minneapolis, Minnesota

Witnesses: John Gribovski
                Mike Gribovski

Craft Type: Nocturnal lights

Primary Color: Various colors, including light green and light yellow

Sound: None

Exhaust: None

Quality of Photos: Sharp and clear

Number of Photos: 1

Type of Camera: Polaroid

Sources: UFO Archives

Reliability: 3

Narrative: Mike Gribovski was outside his house, walking the dog when he glanced up into the night sky in time to see a bright flash of orange. He lost sight of it and then spotted it again, streaking across the northern sky. Gribovski thought at first the object was a meteor, and then he remembered that his camera was sitting in the kitchen. He thought he could get a great picture of what he was seeing.

When Mike ran into the house, his father, John Gribovski, asked him what he wanted and where he thought he was going with the camera. Mike replied that he was going to shoot a picture of a meteor. He ran back outside with his father right behind him.

The object was still in the sky, a faint orange ball now almost due north of the city. When his father saw the object, he told his son not to waste the film. He was convinced the light was too dim to register.

Mike didn't listen. He shot one picture, holding the camera as still as possible. When he looked up, the object was gone and he was sure that he would have nothing on the film. As the picture began to develop, he saw that there was not one object, but five.

John Gribovski said that he had seen the object break up into five pieces. As it did, all of them seemed to accelerate, disappearing in seconds. He described the objects as being dim, with colors ranging from pale yellow to a light green.

Unlike many sightings and pictures taken by teenagers, this one was witnessed by an adult. Although that fact in no way guarantees the authencity of the photograph, it does raise the level of credibility. Given the story and the way it was told, it seems unlikely that it was a hoax. Mike Gribovski's father seemed to be an intelligent man who had no real reason to participate in the fabrication of such a tale. Minneapolis radio news director, Mike Douglas (not to be confused with either the television talk show host or the actor of the same name), interviewed both father and son. He was convinced of their sincerity. He did not believe the case to be a hoax.

El Paso, Texas:

September 24, 1977

Location: Downtown El Paso, Texas

Craft Type: Disc

Primary Color: Black

Sound: None

Exhaust: None

Quality of Photos: Sharp and clear

Number of Photos: 1

Type of Camera: Cosmorex 35 mm

Sources: UFO Archives

Reliability: 6

Narrative: The witness was taking photographs in the El Paso area when he spotted the object moving silently through the sky. He managed to take a single picture of the black, disc-shaped object before it disappeared.

California Disc:

Location: Near Palm Springs, California

Witnesses: Leo Giampietro

Craft Type: Disc

Dimensions: Twenty feet in diameter

Primary Color: Gray

Sound: None

Exhaust: None

Quality of Photos: Sharp and clear

Number of Photos: 1

Type of Camera: Not available

Sources: *APRO Bulletin*, June 1978

Reliability: 5

Narrative: While returning home, driving along Highway 10, Leo Giampietro was searching the skies for military aircraft. Through the left window he spotted a small object traveling through the scattered clouds. At first he was sure that it was a conventional aircraft until he realized that it had no wings or airfoils.

It dawned on him that the object wasn't an airplane. He slammed on the brakes so that he could get a better look at the object. A car rushed by, and then Giampietro grabbed his camera, ran across the highway, and started taking pictures. As had happened in so many other cases, as he began to photograph the UFO, the object stopped momentarily. Then it shot straight up, flashing through the clouds until it disappeared.

The Gulf Breeze Sightings:

NOVEMBER 11, 1987

Location: Gulf Breeze, Florida

Witnesses: Ed Walters
                  Possibly several others

Craft Type: Disc

Dimensions: Thirty feet in diameter

Primary Color: Silver and blue

Craft Description: Top-shaped craft with a row of dark squares and smaller openings across the midsection. There was a bright glowing ring around the bottom.

Sound: None

Exhaust: None

Quality of Photos: Sharp and clear

Number of Photos: 26

Type of Camera: Polaroid

Relation to Other Sightings: Muskogee, Oklahoma (1953); New Castle, Pennsylvania (1967)

Reliability: 2

Narrative: On November 11, 1987, Ed Walters, a self-described prominent businessman and a happily married man with a son and daughter, was working in his home office in the early evening hours. He thought he saw something glowing behind a thirty-foot pine tree in the front yard. He stepped outside to get a better look and saw a top-shaped craft with a row of dark squares and smaller openings across the midsection. There was a bright glowing ring around the bottom.

Realizing that this was something very unusual, Walters returned to his office and grabbed an old Polaroid camera. He stepped back outside and took a photograph as the craft moved from behind the tree. In all, he would take five pictures that night as the UFO, about 150 feet away, continued to drift northeasterly.

Just six days later, Walters visited Duane Cook, editor of the *Gulf Breeze Sentinel.* He showed Cook the pictures but claimed they had been taken by someone else. Walters gave Cook a letter allegedly written by the anonymous photographer, explaining the situation. Two days later, on November 19, 1987, the letter and the pictures were published in the newspaper.

On November 20, as Walters returned home and walked through his doorway, he heard a humming in his ears. At first he hardly noticed it, but it grew in pitch until it was nearly unbearable. He walked through the house, followed by his wife, Frances, and then back outside. According to Walters, the hum was the same as that which he had heard during his first encounter with the UFO. The couple saw nothing in the sky.

After his wife and son left the house, Walters picked up his

274

camera, and walked out the front door. Outside, he said, "I hear you, you bastard." There was a rush of air, and a voice said, "Be calm. Step forward."

High overhead there was a speck of light that fell rapidly toward him. He took a picture of the UFO as it hovered above a power pole. Then there were more voices speaking to him as the UFO shot to the right and Walters took a second picture. About that time, the first voice told him to take a step forward so that he could enter the craft. Walters told the speaker that it and its kind had no right to do what they were doing, and the voice said, "We have the right."

A female voice added, "You must do what they say. They haven't hurt us and we are going back home now."

As the first voice said, "We will come for you now," images of naked women filled his mind. Walters took a third picture. The UFO moved forward and then shot upward into the sky, vanishing almost instantly.

He next saw the UFO on December 2, when he was awakened by the sound of a baby crying. Since there were no babies in either of the neighbors' houses, or his own for that matter, Walters was upset. He then heard the voices, speaking Spanish, and talking about the crying baby. Accompanied by his wife, Walters, carrying a .32 caliber pistol, checked the house and the yard. Out back, he saw the UFO descending rapidly. It stopped briefly about a hundred feet above the pool, then drifted a short distance before stopping again.

Walters retreated to the house to join his wife, who was seeing the craft for the first time. Once again Walters grabbed his Polaroid camera, and took it, along with his pistol, out the door. Near the pool in his backyard, he took another picture, but when the flash went off, he felt exposed. He ran back into the house. From the kitchen, he, along with his wife, saw the UFO vanish.

Back in bed, Walters said he heard the dog bark once, which he characterized as unusual. Walters got up again, and, carrying both his pistol and camera, walked to the French doors, sure that he would see the UFO once again. Instead, when he opened the curtains, he saw, just inches away from him, a four-foot-tall humanoid with big black eyes. It was wearing a helmet with a bit of transparent material at the eye level, which apparently allowed it to see.

Walters, who seemed to have remained calm enough through his other UFO experiences to take multiple photographs of the craft, who disobeyed the commands of the mysterious voices not to photograph the object, forgot about the camera in his hand. He screamed in surprise, jumped back, and tripped. Walters raised his pistol, thinking he would fire it if the creature tried to enter the house, but never thought to take a picture.

Walters finally got to his feet and then struggled with the lock on the door. He put down his pistol and camera. The creature retreated but was no more than twenty feet away. Walters was sure that he could capture it. But, as he opened the door and attempted to step out, he was struck by a blue beam. It seemed that his foot was nailed to the floor. As the beam lifted his leg, Walters grabbed at the side of the doorway for balance. Frances, who had gotten out of bed, grabbed at him and pulled on him. Both saw that the UFO was about fifty feet in the air, above the backyard.

With the UFO hovering over a nearby field, Walters, now free of the blue beam, again grabbed his camera, and shot a picture of the UFO. He didn't manage to photograph the alien being but had the presence of mind to take still another picture of the craft. He saw the object shoot out another blue beam, and Walters believed this was to pick up the creature.

By December 17, Walters had taken seventeen photographs of the object. By the end of December, Walters had figured out that videotape would be more impressive than still photographs. On December 28, he made a videotape that ran just over a minute and a half. According to Walters, his wife, his son, Dan, and his daughter all saw the object.

The next encounter happened on January 12, 1988. While driving on a country road, Walters was hit by two blinding flashes of light that left his arms and hands tingling with "pin pricks" but no other feeling. Five hundred feet in front of him hovered the now-familiar UFO. Walters tried to stop and make a U-turn, but his hands wouldn't obey. He stopped two hundred feet from the object. Although he couldn't drive, he could pick up his camera and take still another picture of the UFO.

As the UFO began to drift toward him, Walters abandoned the truck, trying to crawl under it to hide. Before he could escape, he was hit again by those blue beams and his legs went numb. The UFO was visible, even though he was halfway under the truck. Walters took another picture as a voice told him, "You are in danger. We will not harm you. Come forward." Walters ignored the message.

Five blue beams shot from the craft, leaving five creatures on the ground who began to move toward him. Once again Walters was confronted by alien creatures when he had a camera in hand, but somehow he failed to photograph them. Instead, screaming obscenities, he leaped back into the truck, and drove off. Apparently his hands and legs were working fine by that time.

Over the next month, Walters continued to see UFOs and photograph them. On January 21, he was in communication on a walkie-talkie with Bob Reid, a colleague, who was staked out a block away with a camera. Reid saw the lights that Walters reported, but he identified them as a small aircraft. Walters said that Reid was not looking in the right direction to see the real UFO.

At the end of February, the Mutual UFO Network provided Walters with a special camera that had four lenses to take three-dimensional photographs. The camera produced four negatives for each picture. It should have made it possible to gather a variety of technical information about the object, based on measurements from the negatives. That evening, Walters took more pictures of an object, or at least, took pictures of lights in the distance in the night sky. Frances thought the object was small and close, but Walters thought it was larger and farther away. Strangely, none of those pictures matched the spectacular nature of the other photographs that Walters had taken on other occasions.

On March 8, Walters returned to a Polaroid camera, now using a newer model. Again he took a picture of the UFO, this time hovering about three hundred feet beyond two pine trees. This photo was much better than the pictures of distant lights he had taken with the special, sealed camera, which could have provided precise technical data.

At the beginning of May, Walters, again alone, was in the park with a SRS camera when he heard the faint hum. This time he shouted, "Here I am! I want you out of my life!" As he attempted to photograph the object yet again there was a blinding flash and Walters lost all sensation except for a feeling that he was falling. About an hour later, he regained consciousness. This was the last encounter that Walters reported.

Because of the nature of the case, and the number of sightings, potential corroborating witnesses, and the existence of so many photographs, there were a number of investigations that were launched. Researchers from the J. Allen Hynek Center for UFO Studies, including Robert D. Boyd, were convinced, almost from the beginning, that the case was a hoax. Boyd felt that Walters did not react as someone who had had six months of self-proclaimed horrifying UFO experiences would react. In fact, it was noted by Center investigators, that the only cases in which a witness claimed repeated encounters with multiple photographs were either known hoaxes or strongly suspected to be hoaxes.

On the other hand, the Mutual UFO Network's investigators, including Don Ware and Charles Fannigan, were convinced that this was one of the best cases to have been reported to date. Dr. Bruce Maccabee of the Fund for UFO Research was also convinced, based on his professional examination of the photographs, that Walters was telling the truth and that the pictures showed a real craft from another world.

There were hints about the reality of the case. One of the first to suggest that there was more to the Walters case than what had been published was Tommy Smith. Around the first of January 1988, Smith told family members that he had seen a UFO and showed them a series of pictures he claimed to have taken. But, about a day later, Smith confessed that the photos were part of a prank that Ed Walters, also known as Ed Hanson to those in the Gulf Breeze area, was playing.

According to an investigation conducted by Carol and Rex Salisberry, Smith told his family that Walters had given him the photos and told him to take them to the *Gulf Breeze Sentinel*. There, he was to claim to the editors that he had taken them. He also said that he had seen two UFO models at the Walterses' home and that he had seen Ed Walters photograph one of them. According to the report prepared by the Salisberrys, Smith said that Walters' wife, son, and another teenager named Hank Boland were all involved in the hoax.

Smith told family members who didn't know what to do about this information, but his father, Tom Smith, Sr., asked his law partners and then Gulf Breeze chief of police Jerry Brown what his son should do. They all decided that the best action, at that moment, was no action. They believed, that since many people in Gulf Breeze already knew the pictures were part of a practical joke, the interest in them would die quickly.

Of course, that didn't happen. Interest in the photographs continued to spread with national television audiences having a chance to see them. On June 19, 1988, Gulf Breeze mayor Ed Gray, called a press conference. Tommy Smith's account was substantiated by sworn testimony and independent interviews conducted with the principals.

Smith was given a number of tests in an attempt to verify his veracity. According to the Salisberrys, a recording of one of the interviews had been made. It was the opinion of a number of professionals that the recording could be used in a voice-stress analysis. In a report dated October 10, 1990, Dale Kelly, in a signed statement for the Gulf Breeze chief of police, wrote, "At the request of and under the authority of Chief Jerry Brown of the Gulf Breeze Police Department, I analyzed a tape of a person known only as Chris [Tommy Smith] to me. The subject matter was the taking of photos of 'UFO.s' and if the photos were faked. Based on the test results, it is the opinion of this examiner that 'Chris' was telling the truth when he described how he was told how the photos were faked. In answer to all questions put to 'Chris,' in my opinion he was telling the truth."

In a second report dated October 18, 1990, Ed Halford, in a signed statement for the Gulf Breeze chief of police, wrote, "I ran a test for the chief of police in Gulf Breeze, Fl., to determine the truthfulness of a statement made by a male identified as 'Chris' [Tommy Smith]. . . . In my professional opinion, the answers to all the questions asked of this person were truthful. I used the Mark II Voice Stress Analyzer to arrive at this conclusion."

Of course, when Smith's allegations were printed in the *Pensacola News Journal*, there was a response from the UFO community, especially those who believed the photographs to be real and Ed Walters to be truthful. By this time, a model of the UFO had been found, and to many, it was the smoking gun proving the case a hoax.

According to the massive report prepared by the Salisberrys, after their intense and exhaustive investigation, Craig Myers, a staff writer for the Pensacola newspaper, told of how the model was accidentally found. Walters had sold the house from which he had repeatedly seen the UFO. Myers, according to a statement in the report, went to interview the new owners on June 4, 1990.

Myers wrote, "Because the Menzers live in the house where Walters reports he had encounters with aliens and photographed UFOs, Myers was curious if the Menzers had ever seen anything unusual.

"During the interview Myers asked if they had ever seen or heard anything unusual, found any darkroom materials, models, etc. The Menzers said they had found what may be construed as a UFO model, and loaned it to the *News Journal*. During the next several days the model was used in an exhaustive series of photographic experiments."

Farther down in the statement, Myers wrote, "Using the model we were able to recreate photographs very similar to those Ed Walters printed in his book. Walters and his supporters have stated that the photographs are not the same because most of his UFOs had two rows of windows. However, a second row of 'windows' can easily be recreated by drawing them on the lower portion of the model."

In what is an important point, Myers wrote, "On Saturday, June 9, 1990, *News Journal* Managing Editor Ken Fortenberry interviewed Walters in Fortenberry's office. Metro Editor Joedy Isert and reporter Nathan Dominitz witnessed the interview in which Walters denied any knowledge of the UFO model, but refused to take either a lie-detector test or a voice-stress analysis conducted by independent experts. Walters did, however, sign a sworn statement denying any knowledge of the model. Walters said the model was obviously 'planted' in his former residence by debunkers, and intimated that the government may have been behind the debunking plan."

It would seem, then, with the testimony of Tommy Smith, with the discovery of the model, and the misleading statements made about the case, the only conclusion to be drawn is that the Gulf Breeze photographs and the accompanying story were little more than a hoax. But the supporters had their own version of the events. They insisted that there was a conspiracy to destroy the case. Solid investigation, corroboration from additional witnesses, and the shady background of those suggesting a hoax would prove to believers that there was no hoax.

There were parallel events as well: Walters hadn't just taken pictures of UFOs, but had also taken pictures of ghosts during teenage parties held at his house. These photographs taken during those parties is illustrative of the mind-set of Ed Walters. Again, according to the investigation conducted by the Salisberrys, "They said that Tommy [Smith] was aware of Ed's tricks and even was Ed's accomplice in a stunt. . . . In one instance of the stunt being played, a girl's name (obviously preselected) was also made to mysteriously appear on a board when a match was struck.

According to several witnesses the girl was so scared by this that she ran from the room in tears. (Ed told Charles Flannigan, Rex, and myself that he had chosen Tommy to be in on the stunt because Tommy was so quiet that no one would suspect him of being an accomplice.)"

Salisberry, attempting to corroborate the accounts of these parties, investigated further. She reported, "I asked one young woman if she had ever gone to a party at the Walters' home. Her reply surprised me. 'It was no party; it was a séance! I'm a Christian, and I was offended by what happened there and I never went back again.' She explained that there was a pentagram or star on the floor and that Ed had 3 girls sit in the middle of a circle surrounded by the other guests. Then he read the 23rd Psalm backwards, having the kids recite after him. (Summoning the ghost for the Polaroid pictures of 3 individuals, one of which would be the chosen one. The chosen one would have the ghost in the picture with her.) This girl's brother, who was also present at this party, and some of the others I interviewed verified this. . . . The kids, now all graduated from high school, said that they couldn't figure out how Ed did the pictures or some of his other tricks. Those interviewed considered Ed to be very clever and that he seemed to know a lot of tricks and games. . . . Several of these witnesses who knew Danny [Walters's son] . . . said that they thought it was unusual that Danny never spoke about the numerous UFOs that appeared at his house. . . ."

There is also the testimony of a number of other teenagers who were there when some of the jokes were played or who saw the results of the trick Polaroid pictures. Some were offended by Walters's séances, and others participated in his practical jokes. In other words, there was a great deal of corroboration for the fact that Walters played practical jokes and created pictures of ghosts using a Polaroid camera.

There is very good evidence, then, from a number of people, that Walters used a Polaroid camera in his jokes. He created photographs of ghosts to fool the teenagers at these parties. A Polaroid camera was used to produce evidence that the séance had been real and that there were ghosts in the room. The effects are all the result of double exposures, which demonstrate that Walters knew how to produce them.

Carol Salisberry concluded, "Based on the information given in peer group interviews, it seems that the Walters had a variety of parties and also small group gatherings at their home. The same teens did not always attend the gatherings. All the parties did not involve a mock séance or spooky tricks but there seems to have been several parties in 1986–87 where these things did take place and a ghost photo was taken. Most of the interviewed teens and parents stated that when the UFO pictures first came out in the paper they thought it was just another of Ed's jokes."

All the evidence seems to weigh against Walters. He stands alone against a large number of witnesses to his love of jokes, his ability to fake photographs with Polaroid cameras, and his attempts to induce others to join in the jokes. All of this makes a great circumstantial case suggesting the story is a hoax, but the die-hard believers are always going to say the evidence doesn't prove that beyond a shadow of a doubt. They say find real evidence that the pictures were faked.

That final test came from an examination of the photographs that Ed Walters claimed to have taken over a period of several months. Some had suggested that if one photo was proven to be a hoax, then it could be claimed that all were a hoax. That seems to be a valid theory. After all, if a man is taking photographs of a real object on a number of occasions, what purpose could be served by faking one? Instead of having twenty such pictures, he has only nineteen, which puts him way ahead of everyone except contactees.

One of the Ed Walters's pictures has been proved to be faked. Photograph No. 19, which Walters claims to have taken from his truck, and which clearly shows the hood of the truck, part of the road, and the UFO, is a hoax. The photograph itself also shows a darkened sky, a tree line, and some other detail seen at twilight. Several disinterested photographic analysts have used that picture to prove the hoax.

Early on in the investigation of the photographs, Bob Boyd tried to warn the MUFON investigator, among others, that there were problems with some of Ed Walters's pictures in general and Photograph No. 19 in particular. On March 7, 1988, he wrote, "The photographic evidence reveals certain inconsistencies which cause suspicion. One example is the state highway 191 B photograph [this is Photograph No. 19, which was taken as the object hovered over the highway] of the object a few feet above the road. The reflection below the object on the pavement does not conform to proper physical features consistent with such reflections."

Although this was a serious defect, Dr. Bruce Maccabee, according to the Salisberry study, conducted an investigation. Using a flashlight, Maccabee reported, "By holding a flashlight at various heights above the road and about 200 feet away it was determined that no reflection in the hood [of the truck] appeared until the light was seven or more feet above the road. This is because the front of the hood was bent by a collision in the fall of 1986. . . ." Later, Maccabee revised his measurements, suggesting that it should be set at six feet.

More important, however, Ray Sanford said that he had examined Photograph No. 19 and that it had been "light blasted and enhanced for detail as published in Walters' book." He noticed that he could see the reflection of the tree line on the hood and believed that the reflection from the UFO illumination should also be visible, especially when it is remembered that not only was there a light ring under the UFO, but porthole lights around the center and some sort of light on the top. Even if the light ring was too low to reflect on the hood of the truck, though there is debate about that, those other lights should have been reflected and they were not.

These observations led to the suggestion that another analysis be performed. The best of the various independent analyses was completed by William G. Hyzer, with an assist from his son, Dr.

James B. Hyzer. Their investigation of Photograph No. 19 revealed that "there was no UFO present and the photo is a product of multiple exposure techniques."

To put all this in context without resorting to minor detail that is unimportant, it must be said that a number of experiments were performed on the road. The Salisberrys, among others, using various light sources, distances, and a truck similar to the one owned by Walters, established an "envelope" of distances, heights above the road, and deflections right or left of the truck. Inside the envelope there would have been a reflection on the truck's hood. According to all the information available, the UFO, as well as the lights on it, fell inside the envelope. In other words, given the location, time of day, and evidence as available on Photograph No. 19 itself, there would have been a light reflection on the hood of the truck if a UFO had been hovering over the road. That there was no reflection was the conclusive evidence that Photograph No. 19 was a double exposure and therefore a hoax.

Hyzer, in his report, wrote, "It is this author's professional opinion that the results of this study are conclusive: if the UFO-like object in photograph number 19 had been real, reflections of luminous sources associated with the dome and dome light at the top of the object, would have to be visible in the truck's hood; but they are not."

The conclusions for the case are obvious to all but the true believers. Ed Walters, playing a somewhat-admitted practical joke, found himself the center of attention, and he loved it. The fact that there was nothing to the sightings meant nothing to him. He grabbed the spotlight as quickly as he could and has done everything possible to stay in it. But his story, from the very beginning, was a hoax.

The Arizona Triangle:

MARCH 13, 1997

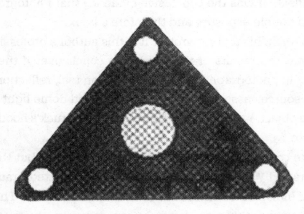

Witnesses in Arizona reported a craft that was similar to this one reported in Belgium.

Location: Arizona

Witnesses: Multiple

Craft Type: Triangular-shaped

Dimensions: Six hundred to eight hundred feet long

Primary Color: Gray

Sound: None

Exhaust: None

Quality of Photos: Sharp and clear and includes videotape

Number of Photos: Numerous

Type of Camera: Various

Relation to Other Sightings: Lubbock Lights (1951); Salem, Massachusetts (1952)

Reliability: 5

Narrative: The first of the Arizona reports began just after eight o'clock on the evening of March 13. A former police officer said that he and his family had seen a strange cluster of lights in the night sky near Paulden, Arizona. It looked to be a V formation of red-orange balls of light. During the next several days, dozens of others reported that they had seen the same thing about the same time.

Similar reports were also made from the Prescott and Prescott Valley areas, about fifty miles north of Phoenix. The witnesses at those locations saw very bright white lights in a triangular-shaped formation. There were those who stated, however, that they, too, had seen red lights in a triangular formation.

Others reported those lights as they moved from northern to southern Arizona, a distance of about two hundred miles in about thirty minutes. Simple calculation showed that the object, objects, or lights, were moving at about four hundred miles an hour. That certainly is not outside the realm of conventional aircraft.

And the speed of the craft might be the explanation for, at the very least, some aspect of the sightings. An amateur astronomer, interviewed by the *New Times* in Phoenix, told the reporters that he had seen the lights in the sky and trained his telescope on them. He said that he could tell they were planes. His account would seem to explain one of the series of sightings in the Phoenix area.

However, there was another set of sightings that took place in and around the Phoenix area, one which resulted in a videotape. These lights were not a group of planes; they seemed to fly in a rigid formation until they began to wink out one at a time. These were the lights, and the videotape, that inspired the investigations of the objects around Phoenix.

The videotape, eventually shown on national television, was analyzed using a number of impressive computer arrays and some very sophisticated software. The conclusion drawn by the televised analysts and supported by their computer data was that the lights seen around Phoenix were not natural and did not exhibit any of the features of manufactured lights. In other words they were unique.

The air force was queried about the videotaped sighting, and its official spokesman said that the air force no longer investigated UFO sightings. Most people in the Phoenix area didn't accept that as an answer. They believed that the air force was hiding the truth.

Others, including private researchers, began collecting the stories of the lights that had moved across Arizona—some of those investigators believing that the sightings were all related. There were those who claimed to have seen a triangular-shaped object and others who claimed to have seen individual lights in a triangular, or V-shaped formation. Some thought the lights were white, others red, and still others saw faded colors in the lights.

Couple these accounts with the hundreds of witnesses to the formation of lights that were videotaped over Phoenix, and an impressive case begins to emerge. Then, eliminate most of those sightings because of air force confirmation of night-flight activities over Arizona on March 13 and the report from the amateur astronomer. Officers at Luke Air Force Base, near Phoenix, reported that they did track a flight of high-flying jets over the Phoenix area at the times in question. That report tends to corroborate the observations of the amateur astronomer. Suddenly everything goes back to the Phoenix videotape.

The analysis of the tape, conducted by Village Labs in Arizona, suggested that the lights were not flares. But the analysis had little scientific legitimacy. It was, in fact, more smoke and mirrors, which looked impressive, but meant nothing.

Other analysis, however, suggested that the lights *were* flares. Although the local air force units said that they had nothing airborne on March 13 that would drop flares, they weren't the only units using the firing ranges. A unit from the Maryland Air National Guard was using the range that night, and their planes did drop flares. The confusion came about because they were not located in Arizona, and it was several days before anyone thought to ask them about flares.

Those at Village Labs, as well as others, said they had observed flare drops on other occasions but these lights didn't match the flares. There was even an impressive bit of computer analysis that broke the wave length of light into a basic spectrum. To some this proved that whatever had been videotaped near Phoenix was not flares.

Others did simpler comparisons. A Phoenix television station videotaped flares and compared them to the Phoenix lights. They could find no differences in the two tapes. Maybe not the most scientific of analysis, but one that did suggest an explanation.

The problem with the flare theory, according to many, was that the lights appeared to be over Phoenix, and not over the range. Certainly the air force wouldn't be dropping flares over the city where they could start fires and kill people.

Another analysis, this one using a bit of videotape taken in the daytime, provided what might be the final explanation. Looking across the valley near Phoenix, there are a number of mountainous ridges. Beyond those mountains are the ranges used for gunnery practice and for dropping flares.

When the original and analysis videotapes are superimposed, so that the mountains are "visible" at night, and the UFO's lights are seen to be hanging in the sky before they wink out, an interesting thing is revealed. The lights, apparently descending very slowly, disappear behind the mountains. They don't burn out or wink out until they fall behind the mountains.

What this information means, clearly, is that the objects were flares, dropped on the range. It was the mountains, invisible at night, and invisible on the videotape, that caused the lights to fade one at a time.

The Phoenix lights, while an interesting case, is one of mistaken identity. The formation flying across the state was high-flying military aircraft, and the lights videotaped over Phoenix were flares.

Part VI:

THE UFO HOAXES

The conventional wisdom is that there are very few hoaxes in the UFO field. Researchers suggest that 90 to 95 percent of all UFO sightings can be explained in the mundane as simple misidentifications of natural phenomena, misidentifications of aircraft or balloons, or as normal things seen under abnormal conditions. Of that 90 to 95 percent, some, maybe as few as 2 percent are hoaxes, according to the researchers. In fact, Project Blue Book officials suggested that there were so few hoaxes, they didn't even deserve their own category.

The truth of the matter is that there have been major hoaxes in the UFO field from the very beginning in 1947. The reason so few of them have been discussed in the UFO literature is that it is very difficult to call someone a liar in print. When a case is labeled as a hoax, those who tell the story are being called, in essence, liars. Most researchers begin to look for other words and other labels to apply to the case. An alternative, if available, is often used instead of the word *hoax*.

In the UFO field there have been a large number of photographs offered as evidence that we have been visited. Unfortunately, the majority of them seem to have been taken by teenage boys, and most of those are hoaxes. This is a fact that is easily verified by a quick examination of those photographs.

It must be noted, however, that many of the UFO researchers have missed those explanations so that pictures exposed as hoaxes surface in UFO books, articles, and on television documentaries as if they are legitimate. It is an area that creates confusion in the general public and journalistic communities and leads those who do not study UFOs, who have a passing interest in them, to believe that there is nothing to them. There is a belief that all of the UFO sighting reports are made by hoaxers, tricksters, and pranksters.

The recent best-seller, *The Day after Roswell*, by retired Lieutenant Colonel Philip J. Corso is a case in point. Corso claimed that during his long military career, he was exposed to the Top-Secret files of various governmental agencies dealing with UFOs. Corso claimed to have an intimate insider's knowledge of what was happening with UFOs, that he had been told about and had seen personally the files about the Roswell UFO crash, and that he

An admitted hoax created by a young man.

The object is a hubcap from an old Ford.

could answer the questions about the crash that had plagued researchers since 1947.

Corso, however, demonstrated that he didn't have access to everything and made a mistake that suggested he might not have access to anything. In the photo section of his book, he published a picture of a UFO over some hills in southern California. He noted that he was never able to confirm the veracity of the "UFO surveillance photos" which he had found in army intelligence files. If Corso was who he said he was, he should have recognized the picture as a hoax. It had been labeled a hoax in the public arena as early as 1966 and the Project Blue Book files had it listed as a fake.

That photograph, according to the editors of a special UFO edition of *Look* magazine, was taken by Guy B. Marquard, Jr., on a mountain road near Riverside, California. Marquard said that it was a hoax, that he was sorry to disillusion people, but he was twenty-one years old at the time and was having some fun. Project Blue Files suggested the object in the photo was the hubcap to a 1930s Ford thrown into the air.

The object looks like a typical airplane.

It would seem that if the vast majority of UFO researchers knew the photograph was a hoax, Corso would have known that as well, if he truly was the insider he claimed to be. Instead, Corso reprinted the photograph as if it was something that had stumped the military investigators.

But Corso isn't alone in his belief that certain photographs reveal the presence of extraterrestrial visitors, which were later proven to be, admitted to be, or shown to be hoaxes. In May 1952, professional photographer Ed Keffel was standing on a cliff near Barra Da Tijuca, Brazil, when he saw what he first believed to be an airplane. The man standing next to him recognized that the craft was something extraordinary and yelled for him to "shoot! Shoot!"

Keffel managed to take five photographs showing an object that was clearly disc-shaped with a dome on the top and a raised ring on the bottom. He was lucky that the maneuvers of the UFO re-

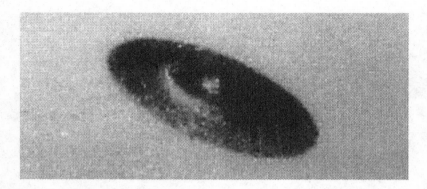

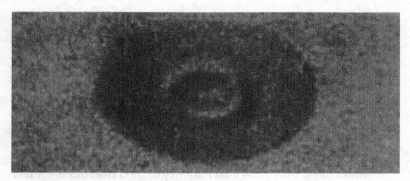

The object turned so that the photographer could take pictures of both the top and bottom.

vealed it to him from all angles. There was no doubt that what he photographed was not an airplane, balloon, or a natural phenomenon.

The Brazilian air force investigated, tracked down an estimated forty witnesses to the sighting, tried to reproduce the pictures with trick photography, and made diagrams of the sighting on-site and of the UFO itself. In the end, according to the report forwarded to the United States by Dr. Olavo Fontes of APRO, they found no evidence of a hoax. At APRO headquarters, the pictures were studied again. APRO researchers found nothing that suggested hoax to them. The pictures, at this point, were termed to be authentic.

The APRO analysis wasn't the last to be performed. During the

Arguments revolved around the shadow on the tallest tree.

University of Colorado study in the late 1960s, the pictures were again analyzed. According to the final report, there was a "glaring internal inconsistency." In the fourth of the five pictures, the object was illuminated from one direction, but the tree in the foreground, specifically a palm tree standing above the other trees, was illuminated from another direction. "This is evidence of a hoax unless there were two suns in the sky," according to the University scientists.

APRO responded to the analysis by insisting that they had known about the problem. According to them, blow-ups of the

The object disappeared in the distance.

photograph showed that one of the palm branches was broken so that it appeared that the tree's trunk was in the shade indicating the "two suns." If not for the broken branch, the trunk would be in the sun. Everything in the picture would then be consistent and the evidence of a hoax was lost.

But even that explanation wasn't the end of the conjecture. People who lived in the area claimed they had seen men with models taking pictures. The Brazilian air force suggested that the people had seen air force officers attempting to duplicate the pictures. They had not seen Keffel and his companion trying to fake the photos.

As it stands today, it seems that these photographs, once considered among the best ever taken, are, in fact, fakes. It is this sort of thing that has plagued UFO researchers from the very beginning of the modern era in 1947.

Keffel wasn't the only man to engage in such a hoax. Paul Villa,

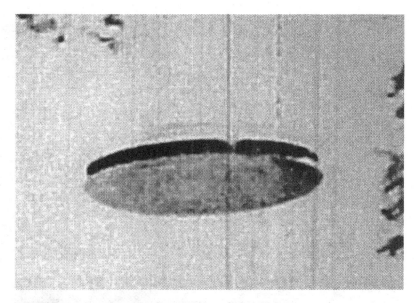

Villa had an unobstructed view of the object.

Jr., of Albuquerque released a number of photographs that he had taken on June 16, 1963. He provided copies of his photographs to the air force for analysis. Not surprisingly, the air force concluded that the pictures were of a small model.

Captain William L. Turner, chief of the Air Force Photo Analysis Division, wrote in his official report to Project Blue Book, "All [Villa's] photographs have a sky background with an unobstructed view of the object. It seems unlikely that anyone photographing a UFO from several angles would have all good, clear unobstructed photographs of the object."

While that might be true, it is also true that Villa might just have been very lucky or even a very good photographer. That, however, doesn't seem to be the case. Turner wrote, "Photograph #7 shows the UFO at close range with a leafless branch on the left side of the print, passing behind the object. Two twigs from this branch are readily visible on the right side of the object and in good alignment

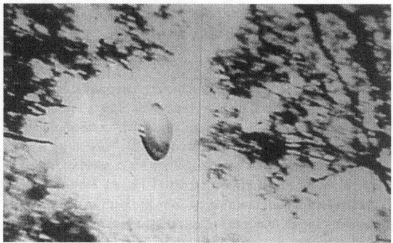

The air force determined the pictures were a hoax, and there has never been a good reason to reject this opinion.

with the main branch. It does not seem possible that these twigs are from the tree on the right which is further away. Therefore, the object is between the branch and the camera. The object is estimated to be 20 inches in diameter and seven inches high."

Turner also noticed one other important fact. He wrote, "In pho-

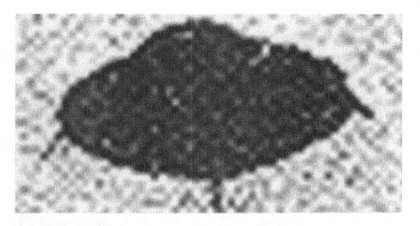

A UFO made from a model kit

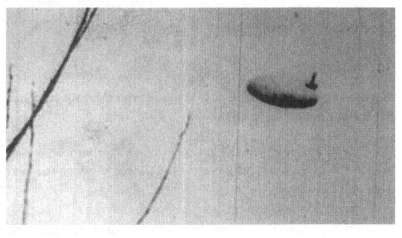

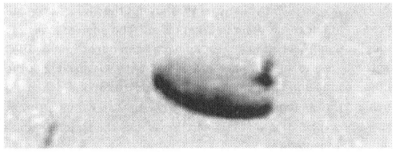

Although originally thought to be authentic, these photographs are now among those admitted as hoaxes.

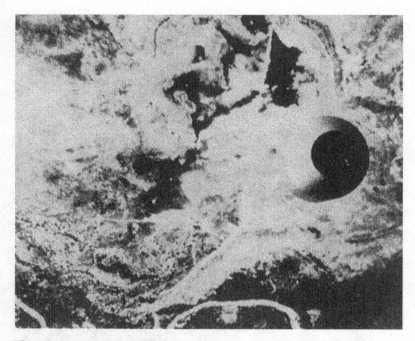

The photographer wanted to fool those who thought him stupid for not believing in UFOs.

tographs #1 and #2 the object appears to be a sharper image than the near and far trees. This indicates the UFO is between the near trees and the camera."

Given all that information, it would seem that the air force had thoroughly destroyed the credibility of the pictures. The question that has been asked by many is, Why accept the air force conclusions here but reject them in other cases? The answer is simply "duplication." The air force results have been duplicated by UFO researchers and civilian photographic experts. It wasn't that the air force presented a complete analysis but that others, when examining the photographs were able to see the same things Turner saw. The explanation was fair, and that is why the air force explanation is accepted.

There are many other pictures that have been published that we now know to be hoaxes. In 1957, for example, Radio Officer

304

Even in the 1950s and 1960s, it was possible to produce sophisticated fakes. Notice the shadows on the ground.

T. Fogel claimed to have photographed a UFO near San Pedro, California. But then he later admitted that he had built the photographed object from a model-airplane kit. APRO published a photograph taken in 1963 that showed an object flying beneath an airplane. The shadows of both could be seen on the ground but it turned out to be a hoax. Two teenagers from Lake St. Clair, Michigan, created a stir with their photographs of a UFO with an antenna on the rear, but later admitted the pictures to be a hoax. One of the very first of the UFO pictures, taken at a steel mill in Hamilton, Ohio, in 1947 is now an admitted hoax.

The list of hoaxes could continue until it was pages long. Today, the hoax-detection problem is getting even worse. Before the advent of computers and various software programs that allow for the manipulation of photographs, it was difficult, but not impossible, to fake good UFO pictures. Something tossed into the air, small models suspended above the ground, objects cut from paper and pasted on the window all contributed to the problem. Analysis by experts could sometimes detect the problems or inconsistencies. It allowed investigators to label a case. If no such inconsistencies were found, it didn't mean that the photographs showed an extraterrestrial craft, only that it couldn't be proven to be a hoax.

Simply stated, in the good old days, researchers had a fighting chance. The picture had to be created physically and mechanically and if they were created in that fashion, there could be something left behind for researchers to find. In today's environment, such is not the case. Any computer and software program can allow the hoaxer to create a negative that can be examined and on which there will be no evidence of manipulation. The hoax-detection job just became that much more difficult.

This difficulty in detecting high-tech hoaxes also explains the problem with videotape. It is why we have mostly ignored videotaped evidence here. It is just too easy to fake a credible videotape with a good computer and very little in the way of video equipment or even expertise. To prove the point we have created just such a tape, but we made the UFO look more like a worm with windows than anything extraterrestrial. We did it so that there could be no confusion about the origin of this tape.

On-site investigations of UFO reports are sometimes very important.

306

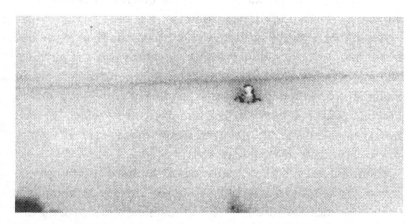

The wire holding up a common cowbell is clearly visible in the second picture.

If, however, we submitted it to any of those specializing in the analysis of videotape, they could digitize it, pixelize it, and analyze it any way they wanted, but they would not be able to tell that we had artificially created that tape. We put the appropriate dialogue on it, making it sound as if we were in awe of what we were seeing. We manipulated the object so that it swooped in, passed behind a tree, and then disappeared in the distance.

The equipment used is not all that expensive, nor is it all that unavailable to the tricksters and the pranksters. Add in the computer software to clean up any problems and then claim tape is the original—no one would be able to tell the difference. The analysis is right back to the credibility of the witness or witnesses. And if they sound sincere, if they have no history of playing jokes and tricks, then there is very little the UFO researcher can do.

So, when studying the photographic evidence of visitor spacecraft, we return to those earlier pictures. Could the witnesses have faked them thirty, forty, or fifty years ago? Certainly. But in that time frame the task was more difficult and the evidence for it often showed on the original negatives. That is why, that long ago, investigators, whether air force officers or civilian researchers, wanted to see the original negatives.

The ideal photographic case would involve multiple witnesses at multiple locations producing both videotape and still pictures. We have often recommended that those with a still camera take a photograph and then move right or left fifty or sixty feet and take a second picture. If possible, the two points from which the pictures were taken should be marked so that precise measurements can be made later by researchers.

What this does is allow investigators to make a stereoscopic view of the object which would provide, on the film, important evidence. The altitude, distance from the camera, and size of the object could all be deduced from a set of photographs made that way.

Now, if there were videotapes of the object, taken by other witnesses in widely separated locations, then corroborative evidence could be collected. It would provide other views of the craft and possibly give additional information about height and speed. It would be a case that would be nearly impossible for the debunkers to destroy because of the physical evidence in the forms of videotapes and stereoscopic pictures. It would end the debate and allow us to move to the next level of investigation.

It would seem, given all the cameras in this country, and now all the videotape cameras available, that we should have evidence like this. Since we don't, it suggests to some that there are no UFOs and spaceships of the visitors.

The situation, if we think about it, is that real UFO sightings are extremely rare. They are usually close to the ground, no more than at a thousand feet or so. That means that only a limited number of people will see them, if they happen to be looking up. It limits the number of available witnesses and the number of cameras.

To argue that meteorites, especially bright ones, are seen by thousands and that meteorites are not only short-lived but also rare, misses the point. The meteorites are usually thirty to forty miles in the air. They can be seen over a wide area. The especially bright ones light the sky, drawing attention to themselves. Often there is a roar associated with the bolides, which also draws attention to them. The UFOs are most often lower, darker, and quieter. Yes, there are a few exceptions, but the vast majority of the cases reflect the lower and quieter component.

So, we are left with a rare and low-flying phenomenon. We are left with photographs, some of which are extremely interesting, but none of which can prove the case of extraterrestrial visitation. The University of Colorado scientists, when studying the McMinnville, Oregon, pictures noted that they found no evidence of a hoax, but they also found that the pictures, by themselves, were insufficient to prove that some UFOs were extraterrestrial craft.

To us, that seems to be a reasonable conclusion because there could be other explanations that do not require the factor of interstellar travel. The fact that we don't have those explanations doesn't mean they don't exist. There might be a natural phenomenon that could account for the pictures. There could be some kind of experimental craft that never reached production that could account for them. We just don't know.

What we do know, however, is that hoaxes, those admitted by the perpetrators and those discovered by analysis by investigators, have plagued the study of UFOs from the very beginning and beyond. The Great Airship of 1897 seems to have been little more than a fleet of hoaxes launched by those tricksters and liars interested in a good story and a good laugh.

Photographic evidence, unless there is a great deal of it from independent witnesses, is never going to provide us with the final solution to the UFO mystery. All they can do is muddy the waters as we learn how many of those photographs were faked by teenagers with too much time on their hands and access to a camera. It seems that nothing has changed since 1897. The people still enjoy a good joke.

Bibliography

Adamski, George. *Inside the Spaceships*. New York: Abelard-Schuman, 1955.

———. *Flying Saucers Farewell*. New York: Abelard-Schuman, 1961.

Asimov, Isaac. *Is Anyone There?* New York: Ace Books, 1967.

ATIC UFO Briefing, April 1952, Project Blue Book Files, National Archives.

"Aurora, Texas, Case, The" *The APRO Bulletin*, (May/June 1973): 1, 3–4.

Barker, Gray. "America's Captured Flying Saucers—The Cover-up of the Century," *UFO Report* (May 1977).

———. "Archives Reveal More Crashed Saucers." *Gray Barker's Newsletter* (March 14, 1982).

Bartholomew, Robert E., and Keith Basterfield. "Abduction States of Consciousness," *International UFO Reporter* 13,2 (March/April 1988): 7–9, 15.

Binder, Otto. *What We Really Know About Flying Saucers*. Greenwich, Conn.: Fawcett Gold Medal, 1967.

———. *Flying Saucers are Watching Us*. New York: Tower, 1968.

———. "The Secret Warehouse of UFO Proof." *UFO Report*, 1976.

Bloecher, Ted. *Report on the UFO Wave of 1947*. Washington, D.C.: Self-published, 1967.

Blum, Howard. *Out There: The Government's Secret Quest for Extraterrestrials.* New York: Simon & Schuster, 1991.

Blum, Ralph, with Judy Blum. *Beyond Earth: Man's Contact with UFOs.* New York: Bantam Books, 1974.

Bontempto, Pat. "Incident at Heligoland." *UFO Universe* (Spring 1989): 18–22.

Bowen, Charles, ed. *The Humanoids.* Chicago: Henry Regency, 1969.

———. ed. *Encounter Cases from Flying Saucer Review.* New York: Signet, 1977.

Brookesmith, Peter. *UFO: The Complete Sightings.* New York: Barnes & Noble, 1995.

Bryan, C. D. B. *Close Encounters of the Fourth Kind: Alien Abduction, UFOs, and the Conference at M.I.T.* New York: Alfred A. Knopf, 1995.

Buckle, Eileen. "Aurora Spaceman—R.I.P.?" *Flying Saucer Review* (July/August 1973): 7–9.

Bullard, Thomas E. *UFO Abductions: The Measure of a Mystery. Vol. 1: Comparative Study of Abduction Reports. Vol. 2: Catalogue of Cases.* Mount Rainier, MD: Fund for UFO Research, 1987.

Cahn, J. P. "The Flying Saucers and the Mysterious Little Men," *True* (September 1952).

———. "Flying Saucer Swindlers." *True* (August 1956).

Canadeo, Anne. *UFOs: The Fact or Fiction Files.* New York: Walker, 1990.

Carey, Thomas J. "The Search for the Archaeologists." *International UFO Reporter* (November/December 1991): 4–9, 21.

Carpenter, John S. "Gerald Anderson: Truth vs. Fiction." *The MUFON UFO Journal* no. 281 (September 1991): 3–7, 12.

———. "Gerald Anderson: Disturbing Revelations." *The MUFON UFO Journal* no. 299 (March 1993): 6–9.

Catoe, Lynn E. *UFOs and Related Subjects: An Annotated Bibliography.* Washington, D.C.: U.S. Government Printing Office, 1969.

Chariton, Wallace O. *The Great Texas Airship Mystery.* Plano, TX: Wordware, 1991.

Chavarria, Hector. "El Caso Puebla." *OVNI:* 10–14.

Clark, Jerome. "The Great Unidentified Airship Scare." *Official UFO* (November, 1976).

———. "The Great Crashed Saucer Debate." *UFO Report* (October 1980): 16–19, 74, 76.

——— "Crashed Saucers—Another View." *Saga's UFO Annual 1981* (1981).

———. *UFOs in the 1980s.* Detroit: Apogee, 1990.

———. "Crash Landings." *Omni* (December 1990): 92–91.

———. "UFO Reporters. (MJ-12)". *Fate* (December 1990).

———. "Airships: Part I." *International UFO Reporter* (January/February 1991): 4–23.

———. "Airships: Part II." *International UFO Reporter* (March/April 1991): 20–23.

———. *The Emergence of a Phenomenon: UFOs from the Beginning Through 1959.* Detroit: Omnigraphics, Inc., 1992.

———. *High Strangeness: UFOs from 1960 Through 1979.* Detroit: Omnigraphics, Inc., 1996.

Cohen, Daniel. *The Great Airship Mystery: A UFO of the 1890s.* New York: Dodd, Mead, 1981.

———. *Encyclopedia of the Strange.* New York: Avon, 1987.

———. *UFOs—The Third Wave.* New York: Evans, 1988.

Creighton, Gordon. "Close Encounters of an Unthinkable and Inadmissible Kind." *Flying Saucer Review* (July/August 1979).

———. "Further Evidence of Retrievals." *Flying Saucer Review* (January 1980).

———. "Continuing Evidence of Retrievals of the Third Kind." *Flying Saucer Review* (January/February 1982).

———. "Top U.S. Scientist Admits Crashed UFOs." *Flying Saucer Review* (October 1985).

Davidson, Leon, ed. *Flying Saucers: An Analysis of Air Force Project Blue Book Special Report No. 14.* Clarksburg, VA.: Saucerian Press, 1971.

Davis, Isabel and Ted Bloecher. *Close Encounters at Kelly and Others of 1955.* Evanston, IL: Center for UFO Studies. 1978.

Davis, Richard. "Results of a Search for Records Concerning the 1947 Crash Near Roswell, New Mexico." Washington, D.C.: U.S. Government Printing Office, 1995.

"The Day a UFO Crashed inside Russia." *UFO Universe* (March 1990): 48–49.

Dennett, Preston. "Project Redlight: Are We Flying the Saucers Too?" *UFO Universe* (May 1990): 39.

Dobbs, D. L. "Crashed Saucers—The Mystery Continues." *UFO Report* (September 1979).

"DoD News Releases and Fact Sheets," Project Blue Book Files, National archives, 1952–1968.

Douglas, J. V. and Henry Lee. "The Fireball of December 9, 1965—Part II." *Royal Astronomical Society of Canada Journal* 62, no. 41.

Druffel, Ann and D. Scott Rogo. *Tujunga Canyon Contacts.* New Jersey: Prentice-Hall, 1980.

Edwards, Frank. *Strange World.* New York: Bantam, 1964.

———. *Flying Saucers—Serious Business.* New York: Bantam, 1966.

———. *Flying Saucers—Here and Now!* New York: Bantam, 1968.

Estes, Russ, producer. *Quality of the Messenger.* (Documentary) Crystal Sky Productions, 1993.

Fawcett, Lawrence and Barry J. Greenwood. *Clear Intent: The Government Cover-up of the UFO Experience.* Englewood Cliffs, NJ: Prentice-Hall, 1984.

Finney, Ben R. and Eric M. Jones. *Interstellar Migration and the Human Experience.* Berkeley: University of California Press, 1985.

First Status Report, Project STORK (Preliminary to Special Report No. 14) April 1952, unpublished.

"Flying Saucers." *Look* (one-time special issue) 1966.

"Flying Saucers Again." *Newsweek,* April 17, 1950: 29.

"Flying Saucers Are Real." *Flying Saucer Review* (January/February 1956): 2–5.

Foster, Tad. Unpublished articles for *Condon Committee Casebook.* 1969.

Fowler, Raymond E. "What about Crashed UFOs?" *Official UFO* (April 1976): 55–57.

———. *The Andreasson Affair.* Englewood Cliffs, NJ: Prentice-Hall 1979.

———. *Casebook of a UFO Investigator.* Englewood Cliffs, NJ: Prentice-Hall, 1981.

———. *The Andreasson Affair, Phase Two.* Englewood Cliffs, NJ: Prentice-Hall, 1982.

———. *The Watchers.* New York: Bantam Books, 1990.

———. *The Allagash Abductions.* Tigard, OR: Wild Flower Press, 1993.

———. *The Watchers II: Exploring UFOs and the Near Death Experience.* Tigard, OR: Wild Flower Press, 1995.

Fry, Daniel W. *The White Sands Incident.* Los Angeles: New Age Publishing, 1954.

Fuller, John G. *The Interrupted Journey.* New York: Dial, 1966.

———. *Incident at Exeter.* New York: G. P. Putnam's Sons, 1966.

———. *Aliens in the Sky.* New York: Berkley Books, 1969.

Gillmor, Daniel S., ed. *Scientific Study of Unidentified Flying Objects.* New York: Bantam Books, 1969.

Goldsmith, Donald. *The Quest for Extraterrestrial Life.* Mill Valley, CA: University Science Books, 1980.

———. *Nemesis.* New York: Berkley Books, 1985.

Good, Timothy. *Above Top Secret.* New York: Morrow, 1988.

———. *The UFO Report.* New York: Avon Books, 1989.

———. *Alien Contact.* New York: Morrow, 1993.

Gordon, Stan. "The Military UFO Retrieval at Kecksburg, Pennsylvania." *Pursuit,* 20, no. 4 (1987): 174–179.

———. "Kecksburg Crash Update." *MUFON UFO Journal* (September 1989).

———. "Kecksburg Crash Update." *MUFON UFO Journal* (October, 1989): 3–5, 9.

———. "After 25 Years, New Facts on the Kecksburg, Pa. UFO

Retrieval Are Revealed." *PASU Data Exchange #15* (December 1990): 1.

Hall, Richard. "Crashed Discs—Maybe," *International UFO Reporter*, 10, no. 4, (July/August 1985).

———. *Uninvited Guests.* Santa Fe, NM: Aurora Press, 1988.

———. ed. *The UFO Evidence.* Washington, D.C.: NICAP, 1964.

UFO (January/February 1991) 30–32.

Haugland, Vern. "AF Denies Recovering Portions of 'Saucers.' " *Albuquerque New Mexican* (March 23, 1954).

Hazard, Catherine. "Did the Air Force Hush Up a Flying Saucer Crash?" *Woman's World*, (February 27, 1990): 10.

Hegt, William H. Noordhoek. "News of Spitzbergen UFO Revealed." *APRG Reporter* (February 1957): 6.

Hopkins, Budd. *Intruders.* New York: Ballantine, 1987.

———. *Missing Time.* New York: Ballantine, 1991.

Huneeus, J. Antonio. "Soviet Scientist Bares Evidence of 2 Objects at Tunguska Blast." *New York City Tribune* (November 30, 1989): 11.

———. "Great Soviet UFO Flap of 1989 Centers on Dalnegorsk Crash." *New York City Tribune* (June 14, 1990).

———. "Spacecraft Shot Out of South African Sky—Alien Survives." *UFO Universe* (July 1990): 38–45, 64–66.

———. "Roswell UFO Crash Update." *UFO Universe* (Winter 1991): 8–13, 52, 57.

Hynek, J. Allen. *The UFO Experience: A Scientific Inquiry.* Chicago: Henry Regency, 1975.

Hynek, J. Allen and Jacques Vallee. *The Edge of Reality.* Chicago: Henry Regency, 1972.

"International Reports: Tale of Captured UFO." *UFO*, 8, no. 3 (1993): 10–11.

Jacobs, David M. *The UFO Controversy in America.* New York: Signet, 1975.

———. *Secret Life.* New York: Fireside Books, 1992.

Jones, William E. and Rebecca D. Minshall. "Aztec, New

Mexico—A Crash Story Reexamined." *International UFO Reporter,* 16, no. 5 (September/ October 1991): 11.

Jung, Carl G. *Flying Saucers: A Modern Myth of Things Seen in the Sky.* New York: Harcourt, Brace, 1959.

Keel, John. *UFOs: Operation Trojan Horse.* New York: G.P. Putnam's Sons, 1970.

———. *Strange Creatures from Space and Time.* New York: Fawcett, 1970.

———. "Now It's No Secret: The Japanese 'Fugo Balloon.'" *UFO* (January/ February 1991): 33–35.

———. *The Complete Guide to Mysterious Beings.* New York: Doubleday, 1994.

Keyhoe, Donald E. *Aliens from Space.* New York: Signet, 1974.

Klass, Philip J. *UFOs Explained.* New York: Random House, 1974.

———. *The Public Deceived.* Buffalo, New York: Prometheus Books, 1983.

———. "Crash of the Crashed Saucer Claim," *Skeptical Enquirer,* 10, no. 3 (Spring 1986).

———. *UFO Abductions: A Dangerous Game.* New York: Prometheus Books, 1989.

———. "Roswell UFO: Coverups and Credulity." *Skeptical Enquirer,* 16, no. 1 (Fall 1991).

Knaack, Marcelle. *Encyclopedia of U.S. Air Force Aircraft and Missile Systems.* Washington, D.C.: Office of Air Force History, 1988.

Lester, Dave. "Kecksburg's UFO Mystery Unsolved," *Greenburg Tribune-Review,* December 8, 1985: A10.

Lore, Gordon, and Harold H. Deneault. *Mysteries of the Skies: UFOs in Perspective.* Englewood Cliff, NJ: Prentice-Hall, 1968.

Lorenzen, Coral and Jim. *Flying Saucers: The Startling Evidence of the Invasion from Outer Space.* New York: Signet, 1966.

———. *Flying Saucer Occupants.* New York: Signet, 1967.

———. *Encounters with UFO Occupants.* New York: Berkley Medallion Books, 1976.

———. *Abducted!* New York: Berkley Medallion Books, 1977.

Mack, John E. *Abduction.* New York: Charles Scribner's Sons, 1994.

McClellan, Mike. "The Flying Saucer Crash of 1948 is a Hoax,"
Official UFO (October 1975): 36–37, 60, 62–64.

McDonald, Bill. "Comparing Descriptions, an Illustrated
Roswell." *UFO* 8, no. 3 (1993): 31–36.

McDonough, Thomas R. *The Search for Extraterrestrial
Intelligence.* New York: Wiley & Sons, 1987.

Menzel, Donald H. and Lyle G. Boyd. *The World of Flying Saucers.*
Garden City, NY: Doubleday, 1963.

Menzel, Donald H. and Ernest H. Taves. *The UFO Enigma.*
Garden City, NY: Doubleday, 1977.

Michel, Aime. *The Truth about Flying Saucers.* New York:
Pyramid: 1967.

National Security Agency. Presidential Documents. Washington,
D.C.: Executive Order 12356, 1982.

NICAP. *The UFO Evidence.* Washington, D.C.: NICAP, 1964.

O'Brien, Mike. "New Witness to San Agustin Crash." *The
MUFON UFO Journal* no. 275 (March 1991): 3–9.

Palmer, Raymond and Kenneth Arnold. *The Coming of the
Saucers.* Amherst, MA, 1952.

Papagiannis, Michael D., ed. *The Search for Extraterrestrial Life:
Recent Developments.* Boston: Kluwer Academic Publishers,
1985.

Peebles, Curtis. *The Moby Dick Project.* Washington, D.C.:
Smithsonian Institution Press, 1991.

———. *Watch the Skies.* New York: Berkley Books, 1995.

Pegues, Etta. *Aurora, Texas: The Town that Might Have Been.*
Newark, TX: Self-published, 1975.

"Press Release—Monkeynaut Baker Is Memorialized," Space and
Rocket Center, Hunstville, AL (December 4, 1984).

Pritchard, Andrea, David Pritchard, John E. Mack, Pam Casey,
and Claudia Yapp, eds. *Alien Discussions: Proceedings of the
Abduction Study Conference.* Cambridge, MA: North
Cambridge, 1994.

"Project Blue Book" (microfilm). National Archives, Washington, D.C.

Prytz, John M. "UFO Crashes" *Flying Saucers* (October 1969): 24–25.

Randle, Kevin D. "Mysterious Clues Left Behind by UFOs," *Saga's UFO Annual* (Summer 1972).

———. *The October Scenario.* Iowa City, Iowa: Middle Coast Publishing, 1988.

———. *The UFO Casebook.* New York: Warner, 1989.

———. *A History of UFO Crashes.* New York: Avon, 1995.

Randle, Kevin D. and Robert Charles Cornett. "Project Blue Book Cover-Up: Pentagon Suppressed UFO Data," *UFO Report,* 2, no. 5 (Fall 1975).

Randle, Kevin D. and Russ Estes. *Faces of the Visitors.* New York: Simon & Schuster, 1997

Randle, Kevin D. and Donald R. Schmitt. *UFO Crash at Roswell.* New York: Avon, 1991.

Randles, Jenny. *The UFO Conspiracy.* New York: Javelin, 1987.

———. *Alien Contacts and Abductions.* New York: Sterling, 1994.

Ring, Kenneth. *The Omega Project: Near Death Experiences, UFO Encounters, and Mind at Large.* New York: William Morrow & Co., 1992.

Ruppelt, Edward J. *The Report on Unidentified Flying Objects.* New York: Ace, 1956.

Sagan, Carl and Thornton Page, eds. *UFOs: Scientific Debate.* New York: Norton, 1974.

Sandreson, Ivan T. "Meteorite-like Object Made a Turn in Cleveland, O. Area," *Omaha World-Herald* (December 15, 1965).

———. "Something Landed in Pennsylvania." *Fate* (March 1966).

———. *Uninvited Visitors.* New York: Cowles, 1967.

———. *Invisible Residents.* New York: World Publishing, 1970.

Saunders, David and R. Roger Harkins. *UFOs? Yes!* New York: New American Library, 1968.

Scully, Frank. "Scully's Scrapbook." *Variety* (October 12, 1949): 61.

————. *Behind the Flying Saucers*. New York: Henry Holt, 1950.

Slate, B. Ann "The Case of the Crippled Flying Saucer." *Saga* (April 1972): 22–25, 64, 66–68, 71, 72.

Smith, Scott. "Q & A: Len Stringfield." *UFO* 6, no. 1, (1991): 20–24.

Special Report No. 14, (Project Blue Book) 1955.

Spencer, John. *The UFO Encyclopedia*. New York: Avon, 1993.

Spencer, John and Hilary Evans. *Phenomenon*. New York: Avon, 1988.

Status Reports, "Grudge—Blue Book, Nos. 1—12."

Steiger, Brad. *Strangers from the Skies*. New York: Award, 1966.

————. *Project Blue Book*. New York: Ballantine, 1976.

————. *Alien Meetings*. New York: Ace Books, 1978.

————. *The UFO Abductors*. New York: Berkley Books, 1988.

Steiger, Brad and Sherry Hanson Steiger. *The Rainbow Conspiracy*. New York: Pinnacle, 1994.

Steinman, William S. and Wendelle C. Stevens. *UFO Crash at Aztec*. Boulder, CO.: Self-published, 1986.

Stone, Clifford E. *UFOs: Let the Evidence Speak for Itself*. California: Self-published, 1991.

Story, Ronald D. *The Encyclopedia of UFOs*. Garden City, NY: Doubleday, 1980.

Strieber, Whitley. *Communion*. New York: Avon, 1988.

————. *Transformation*. New York: Morrow, 1988.

Stringfield, Leonard H. *Situation Red: The UFO Siege!* Garden City, NY: Doubleday, 1977.

————. *UFO Crash/Retrieval Syndrome:* Status Report II. Seguin, TX: MUFON, 1980.

————. *UFO Crash/Retrieval: Amassing the Evidence: Status Report III.* Cincinnati, Ohio: Self-published, 1982.

————. "Roswell & the X-15: UFO Basics," *MUFON UFO Journal* No. 259 (November 1989): 3–7.

————. *UFO Crash/Retrievals: The Inner Sanctum Status Report VI,* Cincinnati, Ohio: Self-published, 1991.

Sturrock, P. A. "UFOs—A Scientific Debate," *Science* 180 (1973): 593.

Sullivan, Walter. *We Are Not Alone.* New York: Signet, 1966.

Swords, Michael D., ed. *Journal of UFO Studies,* New Series, Vol. 4. Chicago: CUFOS, 1993.

Technical Report, "Unidentified Aerial Objects, Project SIGN," February 1949.

Technical Report, "Unidentified Flying Objects, Project GRUDGE," August 1949.

Templeton, David. "The Uninvited," *Pittsburgh Press* (May 19, 1991): 10–15.

U.S. Congress, House Committee on Armed Forces. *Unidentified Flying Objects.* Hearings, 89th Congress, 2nd Session, April 5, 1966. Washington D.C.: U.S. Government Printing Office, 1968.

U.S. Congress Committee on Science and Astronautics. *Symposium on Unidentified Flying Objects.* July 29, 1968, Hearings, Washington, D.C.: U.S. Government Printing Office, 1968.

Vallee, Jacques. *Anatomy of a Phenomenon.* New York: Ace, 1966.

———. *Challenge to Science.* New York: Ace, 1966.

———. *Passport to Magonia.* Chicago: Henry Regnery Co., 1969.

———. *Dimensions.* New York: Ballantine, 1989.

———. *Revelations.* New York: Ballantine, 1991.

"Visitors from Venus," *Time* (January 9, 1950): 49.

Walton, Travis. *The Walton Experience.* New York: Berkley Books, 1978.

———. *Fire in the Sky.* New York: Marlowe & Co., 1996.

Webb, David. *Year of the Humanoids.* Evanston, IL: CUFOS, 1974.

Whiting, Fred. *The Roswell Events.* Mt. Rainier, MD: FUFOR, 1993.

Wilcox, Inez, Personal writings, 1947–1952.

Wilkins, Harold T. *Flying Saucers on the Attack.* New York: Citadel, 1954.

———. *Flying Saucers Uncensored.* New York: Pyramid, 1967.

Index

Printed in the United States
By Bookmasters